TALES OF COMBAT WITH THE 1ST ENGINEER BATTALION IN VIETNAM

MIKE GUARDIA

Copyright 2023 © Mike Guardia

Published by Magnum Books
PO Box 1661
Maple Grove, MN 55311

www.mikeguardia.com

ISBN-13: 979-8-9854285-7-5

All rights reserved, including the right to reproduce this book or any part of this book in any form by any means including digitized forms that can be encoded, stored, and retrieved from any media including computer disks, CD-ROM, computer databases, and network servers without permission of the publisher except for quotes and small passages included in book reviews. Copies of this book may be purchased for educational, business or promotional use. Contact the publisher for pricing.

For Mom, Dad, Marie and Melanie,
and to the memory of Kathryn B. Conner (1925-2023)

Also by Mike Guardia:

Coyote Recon
The Combat Diaries
Hal Moore: A Soldier Once...and Always
Skybreak
Days of Fury
Danger Forward
American Guerrilla
Shadow Commander

Co-authored with LTG Harold G. Moore:

Hal Moore on Leadership: Winning When Outgunned and Outmanned

CONTENTS

Introduction	vii
Chapter 1 – The Past is Prologue	1
Chapter 2 – Diehards	5
Chapter 3 – No Room for Error	28
Chapter 4 – Blood Alley	55
Chapter 5 – Fields of Fire	78
Chapter 6 – Thunder Road	101
Chapter 7 – The Demo Express	116
Epilogue – Requiems & Aftershocks	132
About the Author	143
Select Bibliography	145

INTRODUCTION

August 25, 1966. Specialist Dan Crowley was among the handful of demolition experts assigned to a route clearing mission in support of Operation AMARILLO. By now, these route clearings had become repetitive, and somewhat passé. His equipment load for this mission was the same as it had been numerous times before: a Claymore mine, TNT, C-4 explosives, blasting caps, time fuse, det cord, eighteen rounds for his M-79 grenade launcher, two personal hand grenades, and his Colt .45.

But this would be no ordinary mission.

The Viet Cong had just attacked an American patrol along Highway 16 near Bong Trang. The ensuing firefight became so intense that *three* US infantry battalions were eventually drawn into the melee. Crowley's outfit—Charlie Company, 1st Engineer Battalion—were among the smattering of units thrown together for this impromptu "relief force."

History would call it the Battle of Bong Trang.

In a war marked by controversy and political unrest, Vietnam remains a divisive topic in American history. The war began as an advisory mission to the South Vietnamese government, hoping to roll back the tide of Communism and keep Indochina within the folds of the Free World. However, poor intelligence, unclear objectives, a hostile American public, and a disaffected news media turned the "noble endeavor" of Vietnam into a national nightmare.

During the early years of America's combat mission, however, the collective attitude was much different. Indeed, by late 1966, the Communists were on the defensive, Allied forces were winning every battle, and the goal of "rural pacification" seemed to be within reach. By most estimates, Vietnam was expected to end in a manner similar to Korea: a restoration of the *status quo antebellum*. Although this desired outcome was not to be, the US military nevertheless conducted itself with considerable poise and tactical finesse throughout much of the conflict.

Among the first combat units to deploy to Vietnam was the US Army's 1st Engineer Battalion. In a conflict dominated by airmobile infantry, the combat engineers played a critical role in shaping America's battlefield victories. The engineers' mission was simple, yet comprehensive. They built fortifications, obstacles, tactical bridges; dug defensive positions, set landmines, and performed various types of demolition. In fact, the term "Fire in the Hole!" comes from the verbal expression used by a demolitionist to alert bystanders of an impending explosion. Combat engineers could also fight as infantry whenever ordered.

Fire in the Hole tells the story of Charlie Company, 1st Engineer Battalion during their initial deployment to Vietnam in 1965-66. Told from the perspective of four Charlie Company veterans—Dan Crowley, Larry Blair, Chuck Humphrey, and Jay Franz—this book provides an intimate, no-holds-barred account of the engineers' war in Vietnam.

I am indebted to several individuals who helped make this project a reality. Special thanks are reserved for fellow author Michelle Kaye Malsbury, who arranged my introduction to Mr. Dan Crowley—the Charlie Company veteran whose writings and interviews have formed the backbone of this narrative. Crowley was instrumental in providing photographs, diaries, documents, and the contact information for other Charlie Company veterans. Collectively, these men recount the Vietnam War during its early days—a time when the American public still supported the war… and the soldiers were still optimistic about its outcome.

Since the early 2000s, the American media has given

considerable attention to the passing of our World War II veterans. Comparatively little attention, however, has been given to the rapidly-aging cohort of Vietnam veterans. At this writing, the *youngest* of our Vietnam veterans are now 70 years old. It won't be long before they, too, begin to pass away in record numbers.

Moreover, these veterans have only recently shaken off the stigma associated with their war. Whereas their forefathers from World War II were hailed as heroes, the Vietnam veterans were often protested, spat upon, and called "baby killers." For decades after the war, many of these veterans refused to speak about their service. Others were denied membership in their local veteran organizations. To make matters worse, Hollywood soon jumped on the bandwagon of scorn and mockery. Popular films like *Platoon*; *Full Metal Jacket*; *The Deer Hunter*; *Apocalypse Now*; and *Casualties of War* depicted the Vietnam veteran as malicious, mentally unstable, or a misfit in modern society. Even pop action heroes like Billy Jack, John Rambo, and James Braddock were portrayed as rugged "antiheroes" who struggled with their memories from Vietnam. It wasn't until the release of films like *We Were Soldiers* (2002) that these negative stereotypes began to break.

But just like their World War II counterparts, the Vietnam veterans are no less deserving of the respect and admiration due their service. It is to these veterans—the ones who performed admirably in the field despite being hamstrung by a hostile society—that this book is graciously dedicated.

CHAPTER 1
THE PAST IS PROLOGUE

At the height of the war in Vietnam, Deputy Secretary of Defense Morton Halperin and his assistant Leslie Gelb devised a scheme to compile a multi-volume historical review of US-Vietnam relations since World War II. By analyzing the historical record, Gelb and Halperin hoped to find a solution to the current problem in Vietnam. Perhaps something in the historical data could point them in the direction to finding a peaceful resolution.

What they found would shock them.

The final product was 47 volumes in length, containing nearly 3,000 pages of historical analyses and 4,000 pages of additional government documents. Thirty-six analysts—including military officers, academics, and other federal employees—contributed to the study. Most of their sources came from existing documents within the Pentagon, the State Department, or the field studies brought back from "think tanks" like the RAND Corporation and Booz Allen.

This multi-volume study would later be known as *The Pentagon Papers*.

Of particular interest were the reports from the American intelligence community; especially during the "early years of the war, going all the way back to French colonialism and World War II in particular." In many respects, the pathology of America's involvement read like a modern-day Greek tragedy.

Vietnam had been a French colony[1] from 1887 until the rise of the Viet Minh Independence Movement led by Ho Chi Minh in the 1940s. During World War II, Ho Chi Minh had rescued several downed American pilots in Indochina and, through an OSS detachment, supplied intelligence on Japanese and Vichy French troop movements. All this, he had hoped, would curry favor with the US government and generate sympathy for the anti-French rebellion. In fact, throughout 1945-46, Ho Chi Minh wrote at least eight letters to President Truman and the US State Department, begging for help in the Viet Minh's war of independence against the French. However, there was no record that the US ever replied to Ho Chi Minh's letters or publicly acknowledged their existence.

Indeed, the US had taken a stance of "containment"—fearing that Chinese Communism would spread throughout Asia. Casting their lot with the French, America responded by sending a Military Assistance and Advisory Group (MAAG) to Vietnam in 1950. MAAG was the command group responsible for all US military advisors in foreign countries. It's mission in Vietnam was to supervise the millions of dollars in US equipment being used by the French. By 1953, however, it was clear that the French were losing ground to the Viet Minh.

After the French were defeated at Dien Bien Phu the following year, negotiations at the Geneva Conference separated Vietnam into two political entities: a northern zone, governed by the Communist Viet Minh, and a southern zone, which became the Republic of Vietnam.

At the same time, however, most of the decisions made by the US Government in Vietnam were *against* the advice of the American intelligence community. Indeed, the CIA and State Department analysts had been warning policymakers that the French colonials, Emperor Bao Dai, and South Vietnamese President Ngo Dinh Diem were all weak, and the Communists were stronger than most realized. For example, in August 1954, a national intelligence estimate stated:

---

[1] Part of the greater French Indochina.

"Although it is possible that the French and Vietnamese, even with firm support from the US and other powers, may be able to establish a strong regime in South Vietnam, we believe that the chances for this development are poor and moreover, that the situation is more likely to continue to deteriorate progressively over the next year."

But as *The Pentagon Papers* concluded: "Given the generally bleak appraisals of Diem's prospects, they who made US policy could only have done so by assuming a significant measure of risk." Thus, it raised the question: Why did the US pursue its Vietnam policy despite warnings from its most seasoned intelligence officials?

The answer was the "domino theory."

The National Security Council stated in 1950 that: "The neighboring countries of Thailand and Burma could be expected to fall under Communist domination if Indochina is controlled by a Communist government. The balance of Southeast Asia would be in grave hazard." But, as stated in *The Pentagon Papers*: "The domino theory and its assumptions were never questioned."

Per the Geneva Accords, the Republic of Vietnam was to hold a reunification election in 1956. However, Ngo Dinh Diem cancelled the elections and vowed to stamp out any lingering Communists in the Republic of Vietnam. The Viet Minh operatives who remained in the south (the first incarnation of the Viet Cong) reciprocated by launching a low-level insurgency in 1957. MAAG, meanwhile, stepped in to assist Diem in his anti-communist efforts.

The US advisory mission continued into the early 1960s. But by the fall of 1963, the Viet Cong insurgency had grown to a level that Washington could no longer ignore. Soon, MAAG would be dissolved into the newly-created Military Assistance Command, Vietnam (MACV)[2]—thereby giving the US a wider berth to send conventional forces into Southeast Asia. Almost simultaneously, President Kennedy lost his confidence in Ngo Dinh Diem's ability

---

[2] Pronounced "mack-vee."

to rule South Vietnam. On November 2, 1963, just weeks before Kennedy's own assassination, Ngo Dinh Diem was deposed and murdered in a coup d'état that was sanctioned, if not partially orchestrated by Washington.

That's when the powder keg in Vietnam finally exploded.

In the wake of Diem's assassination, Saigon went through a series of violent coups staged by South Vietnamese generals who took turns being "strongman of the month." At the same time, the Viet Cong continued to grow in the Mekong Delta and began exerting their influence in the Central Highlands and the Coastal Plains. In the middle of it all, South Vietnam's army (Army of the Republic of Vietnam—ARVN)[3] remained poorly-led and largely unmotivated.

In August 1964, Congress passed the infamous Gulf of Tonkin Resolution. The new measure was drafted in response to a naval skirmish involving North Vietnamese boats and the US destroyers *Maddox* and *C. Turner Joy*. Essentially, it gave President Lyndon Johnson the unprecedented authority to use conventional military force in Vietnam without a formal declaration of war. Still, Johnson was confident that he could find a diplomatic solution to the Vietnamese problem.

All that changed, however, on the night of February 15, 1965, when Viet Cong sappers attacked the US airbase at Pleiku. That night, a fed-up Johnson went to his National Security Council and said, "I've had enough of this." The following month, he authorized a systematic bombing campaign and, on March 8, 1965, the first US Marines waded ashore at Danang.

Thus began America's combat mission in Vietnam.

---

[3] Pronounced "arvin."

CHAPTER 2
DIEHARDS

CROWLEY

Dan Crowley

Born in January 1946, Dan Crowley was the second of two sons born to a Merchant Marine officer. Although the US Merchant Marine was a commercial fleet, many of its officers were Naval Reservists or served in unofficial military capacities—transporting troops and materiel as part of naval convoys. Such was the case with Crowley's father, who spent much of World War II crisscrossing the Atlantic in support of the Allied effort. Unfortunately, Dan never saw much of his father; and after his parents divorced at the age of five, he saw his father only twice more thereafter.

Dan's mother soon remarried and, as he recalled, his stepfather Jerry was a positive influence in his life. "My stepdad was in the Army during World War II...in a chemical unit," he said—armed with poisonous gas, and ready to use it only if the Nazis drew "first blood" on the chemical front. Luckily, neither side made extensive use of their biological weapons. And the Allied chemical units in Europe were often relegated to *non-chemical* tasks like prepping incendiaries for flamethrowers or performing smokescreen missions. After the war, Jerry acquired a dairy farm and resort, where he let his stepsons have free range on the property.

As Dan recalled, it was an idyllic life he shared with his older brother, Jack. Their stepfather's property had some densely wooded areas, several lakefront cabins, and a beloved black Labrador Retriever named Duke. In many ways, Dan's childhood was typical of most boys growing up in small town America. He was a member of the local Boy Scout troop, immersed himself in the local variety of team sports, and became quite adept at life on the farm. In fact, he later recalled that working on tractors and other pieces of farm equipment sparked his interest in military engineering.

Perhaps it was inevitable that Dan Crowley would become a demolitions expert, for his interest in pyrotechnics began at a young age. Because his mother and stepfather owned businesses in Florida and Michigan, Dan spent much of his childhood as a "snowbird," travelling between the two states to avoid the northern winters. "Riding in the back of our 1957 Dodge station wagon," he recalled, "on my first trip to Florida, I spotted a billboard that made my heart skip a beat." Indeed, this brightly-colored billboard advertised the local *Stuckey's*—the famous roadside truck stop and trading post known for its Pecan Roll delicacies. But this particular franchise was advertising something more:

"*Cherry Bombs, M-80s, Fireworks.*"

All for sale at bargain basement prices. "My older brother Jack and I had been making homemade fireworks out of shotgun shells, rifle ammo, matchheads, and gas," he admitted, "but Stuckey's had the real deal—boxes and boxes of fireworks, racks of lady fingers, black cats, and silver salutes—you name it, Stuckey's had it. I was in awe!"

But to his mother and stepfather, Dan's adolescent enthusiasm for pyrotechnics seemed to border on pyromania. One day, he and Jack were snooping around one of the buildings on the farm, where they stumbled upon a box of dynamite, fuse, and blasting caps.[4]

"Oh boy! We are going fishing."

Dan and his brother were going to *blast* the fish out of the water.

Taking their friends down to the local river, they tied some dynamite to a rock...and threw it into the spot where the Walleyes ran deep. "When it went off, it was loud and we woke up a lot of people," he said. "It was a mad scramble to gather the fish before someone came looking." Luckily for Dan and his friends, they made off with three bushels and two full sacks of fish before their irate neighbors arrived on the scene.

It wasn't long, however, before Dan's pyrotechnics caught the attention of local law enforcement. "I had been selling M-80s and cherry bombs," he said, "and a lot of things around town were being blown up—mainly mailboxes."

And Dan's customers quickly ratted him out whenever they were caught.

Although no one could fault the young Crowley for his enterprising capitalism, the local parents were none too thrilled by the wasteland of mailboxes. As it turned out, the chief of police was one of his mother's childhood friends. And the constable's advice to Dan's mother was simple and direct:

---

[4] Dan clarified that his stepfather kept a stash of dynamite for clearing beaver dams from the farm's numerous streams. Dan recalled that using dynamite was the only guaranteed method for clearing beaver dams. "You've got to hand it to them though," he said, "the beaver is the Combat Engineer of the rodent world. He can build an abatis, a moat, and a castle, and he has the teeth and tail to defend it."

"Stop with the fireworks!"

Reluctantly, Dan surrendered the last of his inventory. "Even as cool as my stepdad Jerry was, he was over the fireworks, too." Pyrotechnic side hustles notwithstanding, Dan Crowley was a model student. By the end of his high school career, however, he had received his letter indicating that he was now *eligible* for the peacetime draft. After all, it was 1964…and most young men expected to serve a two-year conscription at some point in their lives. To Dan Crowley, however, this letter read like a thinly-veiled recruiting pitch. As if to say: "Join now, and choose your branch of service, or we'll eventually make that choice for you."

At first, Dan thought about joining the Air Force. He had been a Civil Air Patrol cadet and was somewhat familiar with the Air Force culture and ranking system. But, by virtue of being the Air Force, the recruiters were never hard up for prospective recruits. Plus, the Air Force wanted a longer commitment than Crowley was willing to give at the time. His older brother, Jack, was a recent Navy veteran and had served aboard a destroyer. Although Jack had enjoyed the life of a "tin can" sailor, his extended deployment during the Cuban Missile Crisis didn't appeal to the younger Crowley.

Thus, owing to his childhood fascination with pyrotechnics, explosives, and his work on the farm, Dan thought the Army seemed like a good fit. "I was into tractors, combines, plows, and stuff on the farm," he said, "so it seemed like a good thing to get into the combat engineers." Half-jokingly, he added: "And if I learned how to run a bulldozer, I'd have a marketable trade if I got out!"

Enlisting in 1965, Crowley reported to Basic Training at Fort Jackson, South Carolina. Coincidentally, a few of his drill sergeants were former Green Berets who had completed tours in Vietnam during the latter-day MAAG advisory mission. "They were all Airborne [paratrooper] qualified, and they *loved* to run"—a passion they forcibly imposed on their new recruits. Morning endurance runs would go for several miles, and purposely included routes along the roughest pieces of terrain at Fort Jackson.

Most recruits remember their Basic Training experience as a

blur of running, road marches, obstacle courses, drill & ceremony, and savage putdowns from the drill sergeants. For the young Dan Crowley, Basic Training certainly lived up to its reputation. His most memorable moment, however, came during the fifth week of training.

"I caught pneumonia and had a temperature of about 103."

After the drill sergeants took notice of Crowley's ailment, they sent him directly to Moncrief Army Hospital. Interestingly, however, the medical staff broke Dan's fever by giving him an isopropyl alcohol bath. "I was able to get back without being recycled, and I made it through Basic Training."

After Basic, Dan and his fellow engineer recruits were hoarded onto a train bound for Fort Leonard Wood, Missouri. At the time, Leonard Wood was home to the Combat Engineer Training Center; and the post was infamous for its brutal land navigation course. Purported to have some of the toughest orienteering lanes in the Army, Fort Leonard Wood had earned the nickname "Fort Lost in the Woods." Luckily, for Dan Crowley, the orienteering course presented no big challenges; he had learned how to read a map from his time in the Boy Scouts and Civil Air Patrol. His favorite class, not surprisingly, was the demolitions class.

And he marveled at just how intricate the Army's tactical explosives could be.

"I was so excited about it! We had the primer cord; and it looked like a plastic clothesline, about one-quarter inch in diameter. Then there's the C4. It comes in a 20-pound roll and explodes at about 38,000 feet per second!"

"Then, after Fort Leonard Wood," he continued, "they put us on another train." His orders were to go to New York City and board the USNS *Maurice Rose* headed for West Germany.[5] It would be his first glimpse of the "Iron Curtain"—ground

---

[5] Coincidentally, the *Maurice Rose* was the same ship that transported Hal Moore's unit (1-7 Cavalry) to South Vietnam in August 1965, ahead of the infamous Battle of Ia Drang. It was the first major battle between the US Army and North Vietnamese regulars. The battle was later recounted in the bestselling book *We Were Soldiers Once..and Young*, which was subsequently adapted into the 2002 film *We Were Soldiers* starring Mel Gibson as Hal Moore.

zero along the Frontier of Democracy. After the fall of the Third Reich, Germany had been partitioned along ideological lines: East Germany was now a Communist state, while West Germany had become a capitalist republic. Although the Allied occupation officially ended in 1949, the US established a network of permanent bases throughout the West German countryside. These bases, known as "kasernes," were situated near German towns including Frankfurt, Baumholder, Stuttgart, and Bad Kreuznach. Since the start of the Cold War, the units stationed within these kasernes prepared for an anticipated showdown with the Eastern Bloc. Crowley's gaining unit, the 54th Engineer Battalion, was one such unit committed to the defense of Western Europe.

But for now, his biggest obstacle was simply crossing the Atlantic. The *Maurice Rose*, although a venerated workhorse of the US fleet, still embodied all the stereotypes of a World War II troop carrier.

The ship was cramped, poorly-stabilized, and poorly-ventilated.

"There were about 2,000 troops on that boat when we left New York for Bremerhaven," he said, "and they put us in the hold, with five or six bunks stacked all the way to the ceiling!" Indeed, there was barely one foot of clearance between the top bunk and the overhead (ceiling). "There was barely enough room to roll over between bunks," he added, "and I was skinny back then!" he chuckled.

Before boarding the ship, however, Dan remembered his elder brother's advice. Jack strongly cautioned him against claiming the bottom bunk for the voyage.

But to Dan, this seemed counterintuitive.

Wouldn't sleeping on the *bottom* bunk be more convenient?

"No, no, you want the *top* bunk!" said Jack.

"Why?"

"They'll be walking all over you trying to get up into their bunks."

Moreover, Jack warned him that many would become seasick. And if the men in the upper bunks began to vomit, gravity would not be kind to the man on the bottom-most bunk.

As it turned out, his advice was prophetic.

Indeed, most of the 2,000 GIs aboard the *Maurice Rose* got seasick during the voyage. Dan stayed atop the uppermost bunk, but he felt a tinge of sorrow for the bottom dwellers who were pelted with vomit from their colleagues on the upper racks. At one point, the voyage was so rough that the ship's propellers were cresting the surface. "That was the most disgusting boat ride I had ever been on. Oh God, it was nasty!"

Arriving in West Germany, Dan reported to the 54th Engineer Battalion at Wildflecken. The kaserne at Wildflecken was only a "stone's throw" away from the tri-border area separating East Germany, West Germany, and Czechoslovakia. Disembarking from the *Maurice Rose*, Dan found that his train ride into Wildflecken was far better than his voyage across the Atlantic.

Indeed, Dan and his fellow GIs boarded a *milk train*.

As the name implied, this early-departure train accommodated passengers *and* fresh dairy. Crowley remembered that the train stopped at various farms, picking up milk and cream cans along the way.

"We arrived on post on a Friday afternoon," he said. "First thing was to take a shower. The boat [*Maurice Rose*] had *no* fresh water for showers." After his post-voyage detox, however, Dan made his way to the Enlisted Man's (EM) Club for a fresh order of beer and hamburgers. Dan had a table to himself…until two plainclothes men asked to join him.

"Sure," he replied, thinking these men were fellow GIs.

"Then they started talking to each other in German."

Impressed by their linguistic skills, Dan remarked that the pair spoke excellent German.

"That's because we are German!" they laughed.

Both men were soldiers in the West German Army—the *Bundeswehr*. In fact, the *Bundeswehr* shared many of its facilities with American forces across Germany.

Despite this ongoing partnership, however, NATO's combat readiness had been in a steady state of decline. Beginning in the 1950s, President Eisenhower shifted America's combat stance to one of nuclear deterrence. Strategic retaliation was the rule of the day, and there was some speculation whether ground forces would remain relevant in the forthcoming "Nuclear Age." As a result, US Army Europe's training, readiness, and personnel system began to suffer as more defense dollars were diverted into the growing arsenal of nuclear weapons.

Dan Crowley saw the downstream effects of these policies when he reported to the 54th Engineer Battalion. "The unit was understrength," he recalled, "and being a PFC [Private First Class], I pulled a lot of guard duty and KP [Kitchen Police]." Considering the unit's manpower shortages (and their low standing within the budgetary priorities), "field training" was little more than an afterthought. "However, we did build a float bridge across the Rhine River that could carry a 60-ton tank." Dan also recalled the frequent number of alert drills, most of which occurred during the pre-dawn hours. It was a fast-paced choreography of: "Drop everything; grab your duffle bag; draw your weapon; get on the truck; and go hide out in the forest for a few days," he said. These alert drills tested how quickly NATO's frontline forces could respond to a hypothetical Soviet invasion.

But while these European-based units limped forward under their draconian budgets, the influx of money and personnel into Vietnam continued unabated. To this point, Dan's only knowledge about Vietnam had come from the occasional stories he'd read in the *Stars & Stripes* newspaper. But it sounded like an important mission…and one with a greater sense of urgency than standing post along the Iron Curtain. "In September 1965," he said, "I, along with six others, volunteered for duty in Vietnam."

Three months later, while standing guard in three feet of German snow, Dan Crowley was approached by the Sergeant of the Guard.

"You are relieved! You have orders for Nam! Get packing!"

Crowley's new orders assigned him to the 1st Engineer Battalion. The unit had been in Vietnam for the past few months; and Crowley would be joining them as an individual replacement.

Per his orders: "I am to report to Oakland Army Terminal on

the 25th of December." However, Dan was allocated only two weeks of transient leave, meaning he wouldn't be able to enjoy Christmas with his family. "Mom was pissed that I had to report on Christmas Day."

Arriving in Oakland on the morning of Christmas Eve, Dan wasn't surprised to see that nearly everything was closed for the holidays. Because the shuttle bus wasn't running, Dan and three of his comrades hailed a taxi for the Army Terminal.

"It's 9:00 AM and leave isn't up until 11:59 PM—15 hours to kill."

When they arrived at the Replacement Center, Dan and his fellow GIs notice a hand-written note taped to the door.

Sign in and go to the barracks.

Dan was not amused.

"Screw signing in," he said. "I have 15 hours of leave left!"

He did, however, want a place to keep his duffle bag in the meantime. Thus, he ambled over the barracks, and found an empty wall locker to store his personal gear. With fifteen hours of discretionary time, he planned to get some rest, grab a beer, and perhaps see a movie.

But his plans abruptly changed when a group of fellow Privates stumbled into the barracks.

"They were dirty and smelled bad"—the obvious victims of a work detail.

"They tell us a sergeant will be looking for bodies to clean grease traps."

Having done several rounds of KP in Germany, Dan was in *no* mood for cleaning grease traps. Luckily, he had to use the latrine…and Oakland was one of the few Army posts whose latrines had privacy stalls.

"I was about to leave when I heard the barracks door open."

Dan cringed as he heard the sergeant-in-charge barking at the new arrivals to change into their fatigues. Apparently, this sergeant had just found his new grease-cleaning team. The sergeant already knew these soldiers by name because "they had signed in," said Crowley—"I didn't."

Dan figured that this manic sergeant would soon begin

prowling the premises…looking for extra bodies. "So, I stand on the toilet so my feet can't be seen. The sergeant checks the latrine and leaves. A minute later, I'm right behind him."

Escaping from the barracks undetected, Dan flagged down the same taxi that had taken him from the airport—"he takes me into the strip in Oakland." The city buses were running, so Dan opted for an impromptu sightseeing tour. Of course, being Christmas Eve, most businesses were closed—"lots of open bars and strip clubs, though," he added.

Although he consumed no alcohol that night, he nevertheless took a seat at the local bar. Dan Crowley was 19 years old at the time. And although many states had set their minimum drinking age to 18, California still imposed a 21-year age restriction.

None of this would have been problematic…if not for an overzealous Marine Corps MP. Because Oakland had a high concentration of military personnel, the Navy's Shore Patrol and Marine Military Police made frequent rounds at the local bars and strip clubs, looking for errant comrades who had strayed off the "straight and narrow."

As Dan readied himself to leave, the hulking MP grabbed him off his barstool and flung him outside. Dazed, confused, and now agitated, Crowley produced his leave paperwork, showing the MP that he had until 11:59 to sign in at the Replacement Center.

But this MP wasn't concerned with Crowley's timeline.

Rather, he had pegged Crowley for being "too young" to visit the bar.

And although Crowley was stone cold sober (and had had nothing to drink), the mere presence of a 19-year-old soldier in an alcohol establishment was enough to warrant the Marine's wrath. Within the hour, Crowley was handcuffed and horded into the Navy brig.

"It took two days for the Army MPs to come get me," he recalled. All the while, Crowley was certain that he had broken no law…but the commandant of the Replacement Center didn't see it that way. As the MPs ushered Crowley into the captain's office, he knew it would be an unpleasant meeting. "He gave me an Article 15 [an administrative, non-judicial punishment], fined

$50, and an ass-chewing." All told, Dan was just happy to keep his rank of Private First Class. Indeed, most Article 15s for enlisted men ended with a reduction of rank.

"Next stop—shots and paperwork. I grab my duffel bag, which had been looted while I was in the Brig." Some thoughtful comrades had taken Dan's shaving kit and his extra pair of boots. "Then, I'm on the first plane out of Travis Air Force base headed for Vietnam—all within three hours of getting out of the Navy Brig—talk about getting the Bum's Rush!"

On December 30, 1965, Dan Crowley touched down at the Tan Son Nhut Air Base near Saigon. His next stop would be Di An, home to the 1st Infantry Division Headquarters, and the 1st Engineer Battalion. Admittedly, however, Dan felt a bit uneasy as he beheld his upcoming transport—"a bus with chicken wire on the windows to keep out the hand grenades."

BLAIR

Captain Larry Blair (seated)

Over the course of his childhood, Captain Larry Blair grew up on *three* different Indian Reservations. At the time of his birth in 1938, his parents resided at the Fort Belknap Indian Reservation in Northern Montana. But considering that there were no medical facilities on the Reservation, Mr. and Mrs. Blair had to travel fifty miles west to Havre, Montana for their son's prenatal care and delivery.

Despite the confines of the Reservation, however, Larry was still a child of the times. Indeed, World War II and Korea were *the* defining events of his young life. Although he was barely a grade schooler during the Second World War, he vividly recalled the rationing system and the restrictions on everyday goods. Some of his neighbors were drafted off the Reservation and went overseas to fight, while others took jobs in the growing defense industry. Some families planted "Victory Gardens," growing essential foods to supplement their rations and reduce pressure on the national food system. Others donated their scrap metal to the growing number of "Victory Drives," all of which collected material to be recycled into weapons of war.

Ironically, World War II became a shining moment for the Native American community at large. Nearly 25,000 Native Americans served during the war; and they were not forced into segregated units. Most notably, the Navajo Code Talkers used their native language as a radio code for US forces in the Pacific Theater. This Navajo-based cryptography was so effective that the Japanese were never able to decipher it. Another notable example was Private First Class Ira Hayes—a Pima Indian who was among the six flag raisers on Mount Suribachi during the Battle of Iwo Jima.

Larry's family included a number of heroic men who served with distinction. One of his cousins, for example, was a top gunner aboard a B-17 Flying Fortress. The life expectancy of a B-17 crewman was among the lowest of any combat personnel during World War II. "He was shot down over Germany," said Larry, "and spent about a year and a half in a prisoner-of-war camp." Yet, these tales of survival from his wartime relatives had sparked his interest in the military.

"I was in high school when the Korean War was going on," he continued, "and I was very interested in that." Indeed, he spent much of his leisure time reading the news and any special-interest publications featuring military tactics, weapons, and operational reports from the frontlines. Thus, it came as no surprise when the young high school graduate set his sights on a military career.

At first, however, Larry wanted to join the Marines.

"Four, or five, of my high school buddies decided to join the Marines together," he recalled. The Marine Corps was offering an incentive program wherein a potential recruit could enlist with a friend, and those friends would stay together through Boot Camp and into their occupational training.

"I really wanted to go into the Marine Corps with these

guys," he said.

But Larry's father offered him another proposition.

"If you go to college for one year," said the elder Blair, "then you can do anything you want. You can join the Marines, you can do whatever you want to do, but just go to college for one year."

Half-jokingly, Larry admitted that his father's proposition was a thinly-veiled attempt to make him forget about the Marines. "I'd find the good life...drinking beer and chasing skirts, and I'd forget all about the Marine Corps. And he was right!" said Larry.

But his designs on the military weren't done yet.

Larry enrolled at the South Dakota School of Mines and Technology, which was a "land grant" university. "Being a land grant college," said Larry, "in those years, every freshman and sophomore had to be in the ROTC program. The last two years were optional." However, those who signed up for the advanced-level ROTC courses had to accept an officer's commission upon graduation. An ROTC graduate could be granted either a Regular Army commission (meaning they were assessed as full-time Active Duty officers), or a Reserve commission (serving part-time in the Army Reserve or National Guard).

"And even though I didn't know for sure what I wanted to do [as an officer], I figured I might as well get a Regular Army commission instead of Reserve commission,"—since Regular Army officers tended to get their top choice of assignments. With an engineering degree in hand, Larry saw his best opportunities in the Army Corps of Engineers.

In the summer of 1961, as a newly-commissioned officer, Larry Blair reported to Fort Belvoir, Virginia for the Engineer Officer Basic Course (EOBC). Whereas the enlisted combat engineers (like Dan Crowley) trained at Fort Leonard Wood, the officer corps received its training at Fort Belvoir, nestled on the banks of the Potomac River, twenty miles south of DC.

One week before Larry's graduation, however, the Berlin Crisis broke.

This heated standoff ultimately led to the construction of the Berlin Wall. It was an event memorialized by the staredown between American and Soviet tank crews at Checkpoint Charlie. And coincidentally, Larry Blair had just been assigned to West Germany for his first duty station.

He had expected to arrive in Germany after a few months of follow-on training, but the exigencies of the Berlin Crisis had just accelerated his arrival date. "I had just gotten married," he said. "So, I had to ship my family back to South Dakota." Indeed, given the situation in East Berlin, all US military dependents were being evacuated, and all incoming dependents were told to go elsewhere.

Thus, Larry Blair had suddenly become a "geographical bachelor."

As Larry recalled: "By the time I got to Germany, we were overstrength almost everywhere. They had canceled the orders for everybody that was coming home, and kept them there…but they were shipping their dependents home. We really thought we were going to war with Russia. But by a year later, the crisis was over and they started shipping a lot of guys home who had been extended. And suddenly, where we had a company with five or six officers, now we were down to two."

As the Berlin Crisis abated, Larry began learning the art of tactical leadership along the Iron Curtain.

Admittedly, it was a *steep* learning curve.

The personnel shakeup had rendered his engineer company without a commander. "My first company commander was a World War II veteran," he said. "He'd been a Lieutenant Colonel who'd been [involuntarily reduced] to Captain"—a fate that befell many World War II era officers. Wartime needs had accelerated the promotions of many. Some officers who had entered the war as green lieutenants in 1941, had become field grade officers by 1945. But as the Army tried to normalize and re-balance its postwar ranks, many of these same officers were reduced by two or three pay grades. "And so, this company commander was ready to retire," Larry continued. "He wanted no part of any other war. He just wanted to go home, which he did. But when he left, and then the other senior lieutenants left…we ended up with two second lieutenants in the company. So here I was, with about 10 months in the Army...I hadn't even been a platoon leader for a year, and they made me a company commander. You talk about lost in the woods!"

Luckily, his company had a fine cadre of sergeants. "These non-commission officers [NCOs; the official blanket title given to all sergeants] were really the guys who taught us what needed to be done. They really did a good job." Many of them were battle-hardened veterans of World War II and Korea. "My first platoon sergeant was a veteran of the D-Day landings. He was a tough nut. But, also very compassionate." In this top-heavy company of decorated war veterans, the young Blair was witnessing the best example of Officer-NCO dynamics at the small-unit level.

"By 1965, when I was due to come home, and go to the Advanced Course at Fort Belvoir [teaching young Engineer captains the fundamentals of company command and higher-level staff work] Vietnam was already heating up. So, they diverted me from the Advanced Course and sent me over to the newly-formed Engineer Officer Candidate School regiment."

Indeed, the Army had just created an *Engineer-specific* Officer Candidate program.

Officer Candidate School (OCS) was one of the alternate ways to obtain a commission in the US Army. Known as "90-Day Wonders," these officer candidates were often drawn from the enlisted ranks and received a crash course in small-unit leadership. Like their West Point and ROTC comrades, OCS graduates attended their branch-specific Officer Basic Courses upon commissioning. But now, the Army had decided to recruit Engineer officer candidates, and put them through an accelerated training program that combined traditional OCS with combat engineering topics. "So, I got to be a company commander of the first class to graduate under the Engineer OCS program during Vietnam."

There were sixty candidates in the inaugural class of Engineer OCS. "We split them into two platoons of thirty candidates each. These guys had either volunteered for OCS, or somebody had convinced them that they should go." Although these candidates were motivated, few of them completed OCS. Within one 30-man platoon, for example, only *eight* candidates graduated.

Engineer OCS was just that difficult.

"Of those eight that graduated," said Larry, "they all got together on Graduation Day and bought me a Randall survival knife…it was engraved with their class information on it…and they gave that to me because, by that time, they knew I was going to Vietnam." Indeed, a few weeks prior, Larry had received a call from his career management office. As it turned out, the 1st Engineer Battalion was currently deployed to Vietnam…and they had just submitted an emergency request to fill a company command position.

"I jumped at the chance!"

Years later, Larry took special note of the Randall knife that his OCS students had gifted him. "Of those eight guys that graduated from OCS, two of them got killed in Vietnam. So, that knife has special meaning to me."

HUMPHREY

Chuck Humphrey

Much like Larry Blair, Chuck Humphrey was born and raised in the American Northwest. "I was born in Valley City, North Dakota"—one of three children born to a family of multiple-generation farmers. Growing up on a grain and livestock farm, Chuck recalled that his family's acreage had all the trappings of a 1950s-era Dakota farmstead: "We had milk cows; we had a team of horses; pigs, chickens, sheep; and we harvested wheat, barley, oats…and we had some corn."

Chuck was born in December 1942, a time when the country was mobilizing its farmers to produce more crops for the war effort. Thus, Chuck's father was one of many full-time farmers to receive a draft deferment, owing to the agricultural needs of wartime production. Many of Chuck's uncles, however, fought in World War II…and one saw an additional tour of duty in Korea.

Still, the thought of military service hadn't crossed his mind until he enrolled at North Dakota State University (NDSU) in the Fall of 1960. He was pursuing a degree in agricultural science; but NDSU was also a "land grant" college, meaning that every male student had to enroll in two years of ROTC. After his first few days in the program, however, Chuck discovered that he enjoyed the military lifestyle. In many ways, the fieldcraft reminded him of life on the farm. He then decided to pursue an Active Duty commission while completing his agricultural degree.

As Chuck recalled, it was an interesting time to be an Army ROTC cadet. During his four years at NDSU (1960-64), he had seen newsreels covering the Bay of Pigs Invasion; the Berlin Crisis; the Cuban Missile Crisis; Civil Rights protests; and, tragically, the assassination of John F. Kennedy. Indeed, the 1960s were shaping up to be a turbulent decade.

Like many ROTC programs, the cadre at NDSU included a handful of Korean War veterans—all of whom took great strides to inspire their cadets, and instruct them in the ways of combat leadership. Their service in Korea, however, had made them wary of the "Red Menace;" and they warned their cadets that the ongoing showdown between Communism and the Free World was far from over.

Graduating from NDSU in May 1964, Chuck accepted his commission as a lieutenant in the Corps of Engineers. Reporting to Fort Belvoir for EOBC, he was impressed by the depth of the curriculum. "Topics that were taught in the Engineer Officer Basic Course," he said, "included some of what I expected: construction operations, road design, building bridges, and all that. But we also had a class called 'Atomic Demolition Munitions.'" These small nuclear devices were often called "nuclear land mines"—and they were designed to be exploded forward of the battle area, to block an enemy's advance.

Following EOBC, Chuck arrived at Fort Riley, Kansas for his first operational assignment. It was March 1965, and he had been ordered to the 1st Infantry Division.

"I was assigned to this 1st Engineer Battalion."

No surprises here, he thought.

The 1st Engineer Battalion was, after all, the Division's organic engineer unit.

But the prospect of going to Vietnam was so remote, that he didn't even consider it a possibility. Despite the high-profile Gulf of Tonkin Resolution, Vietnam was still a low-intensity mission—one in which the combat engineers would presumably play no role.

"There's twenty-seven officers in an engineer battalion," said Chuck, "one lieutenant colonel; two majors; a whole bunch of captains," and more than a dozen lieutenants. In fact, most of the lieutenants assigned to the 1st Engineer Battalion were new arrivals—graduates of the most recent EOBC classes. "So, we were brand new Second Lieutenants. Two thirds of us were newlyweds"—including Chuck—"and the other third were unmarried at that time." The single officers were allowed to stay in the on-post Bachelor Officer Quarters.

However, there was no Family Housing available for the young married officers.

"So, we were expected to find a place to live within commuting distance of the military post," he said. "And I think five of us found ourselves renting apartments in Manhattan, Kansas—where Kansas State University is—thirty miles east of Fort Riley."

To ease the financial strain of living off-post, Chuck and his friends formed a carpool. "We traded off," he said, "five of us going in one car for a week at a time. And we'd get up pretty darn early, because we would have to be at Fort Riley by 6:00 AM each morning for physical training."

And the young Lieutenant Humphrey believed that this would be the extent of his military life. He and his fellow junior officers believed that they would be "spending our two years as members of that battalion; members of the First Infantry Division, being involved in training and operations right there at Fort Riley, Kansas."

If a war erupted, it would surely happen in Western Europe long before it happened in Vietnam.

FRANZ

Born in 1937, Jay Franz was the son of a New Deal family. His father had worked in the Civilian Conservation Corps (CCC) during the darkest days of the Great Depression, and later obtained a commission in the Army Air Corps on the eve of World War II. Like most military families, Jay and his siblings lived a fairly itinerant lifestyle. "We moved from Pennsylvania to Virginia, to Kentucky, to Texas, to California. While we were in California, my dad got orders to go overseas" -wherein he would serve in the China-Burma-India (CBI) Theater. The CBI was, in many respects, the forgotten theater of World War II. While most of the headlines (and defense dollars) went to the European or Pacific Theaters, the CBI contingent typically got the leftovers. Still, the CBI produced such heroes as Merrill's Marauders and the Flying Tigers.

When the war ended, the elder Franz wanted to stay in the military, but he was involuntarily separated as the military rapidly demobilized and trimmed its postwar ranks. But in 1948, the newly-created United States Air Force offered him the chance to return to active duty as an Air Intelligence Officer. Thus, it wasn't long before Jay's father was back in uniform, proudly serving at Lowry Air Force Base in Denver, Colorado.

Despite his military pedigree, Jay had never seriously considered a career in uniform. "I always did fairly well in school...and people seemed to assume that I'd go to college," he said. "But I had no idea where I wanted to go. And I don't think my folks had any idea how they'd pay for it." His father, however, had an idea:

"What would you think about going to West Point?"

"Why not?" said Jay.

He had a passing familiarity with the academy, but didn't know much about West Point beyond that its graduates served as Army officers.

The application process included a prerequisite nomination from a congressman or Senator. At the time, members of Congress allocated these nominations based on the results of a competitive civil service examination. When Jay sat for the exam, he was convinced that his chances of getting a nomination to West Point were higher because most of his contemporaries wanted appointments to the brand-new Air Force Academy. That year, 1954, was the inaugural year that the Air Force Academy began accepting applications. Jay recalled that less than a handful of the candidates were seeking nominations to West Point.

Reporting to West Point in the summer of 1955, Jay recalled that: "I had an interesting four years there." As a member of the Class of 1959, Jay wanted to commission into the Corps of Engineers. At the time, Engineers was the preferred branch for many West Point graduates. "Branch assignments at West Point were based on class rank," said Jay. Thus, the high-ranking cadets got their branch of choice before anyone else. And given the panache of the Engineer branch, it was nearly always the first branch to be filled up.

"I wanted to go into the Corps of Engineers," he said, "but I didn't make the cutoff."

He did, however, make the cutoff for his second choice: Infantry.

Thus, after completing the Infantry Officer Basic Course (and Airborne School) at Fort Benning, Georgia, he reported to Fort Richardson, Alaska.

Like most young lieutenants, his first duty assignment was as a platoon leader—but not in the capacity he had expected. "My platoon was the honor guard," he said - the official drill-and-ceremony soldiers for US Army Alaska. "We might be out playing in the snow twenty miles from Fort Richardson," he continued. "Then, if a VIP decided to come down, we'd get an advanced notice…usually a couple hours before the VIP's arrival…then we'd head back to post, get our good-looking uniforms on, and go down to greet the VIP." Depending on their training requirements, Franz's platoon would often go back to the field immediately following the dog-and-pony show.

All in all, it was exhausting work for any young lieutenant.

"Anyhow, after two years of being commissioned," he said, "I was able to put in for a branch transfer." He could now pursue his dream branch. "I transferred to the Corps of Engineers, and went to the 562d Engineer Company, which was two buildings up the road from where I had been!"

While learning the art of military engineering, however, Jay took heed of President Kennedy's request for volunteers in Vietnam. "And this was before anybody had ever heard of Vietnam," said Jay. Still, it was an operational deployment, and Jay didn't want to pass up the opportunity.

It turned out that Jay Franz was among the first officers in Alaska to go to Vietnam. It was 1962, and Jay arrived in Saigon as an "advisor" to a South Vietnamese engineer battalion. Because he had been a recent branch transfer, and received on-the-job engineer training at Fort Richardson, "I never went to the Engineer Officer Basic Course," he admitted. But by the time Jay's "educational gap" came to the attention of the top brass in Saigon, he had already been advising the ARVN engineers for several weeks.

His steep learning curve, and solid performance, however, warranted consideration to give him "constructive credit" equivalency for the Officer Basic Course. "Not too long after that," he said, "I got a letter from the Chief of Engineers saying: 'Upon completion of this tour, you'll be favorably considered for Advanced Schooling.'" Essentially, the Army would allow him to attend graduate school, then have a follow-on assignment to the Engineer Officer Advanced Course (EOAC)—learning the fundamentals of higher-level staff work and company command. "And that's how it worked out. I went from Vietnam to the University of Illinois, and got a Master's in Civil Engineering," then EOAC at Fort Belvoir.

In the spring of 1965, Jay Franz arrived at Fort Riley, whereupon he assumed command of Charlie Company, 1st Engineer Battalion.

CHAPTER 3
NO ROOM FOR ERROR

On Memorial Day 1965, Chuck Humphrey received an unexpected call from his mother. For the past few months, she had been following the news in Southeast Asia. Earlier that March, the Marines had established a beachhead at Danang. Two months later, the US Army's 173d Airborne Brigade arrived in Saigon. From these developments, Mrs. Humphrey was certain that the military commitment in Vietnam was growing.

"Do you know what's going on," she asked, "over in this place called Vietnam?"

Chuck laughed.

"Mom, I don't," he replied. "But I can tell you…I'm the assistant maintenance officer in our engineer battalion. We have to produce a report every month on the serviceability of our equipment." Given the current status of their inventory (due to their lower ranking within the Army's annual budget), Chuck was certain that it would take nearly *twelve months* for the unit to achieve a "deployable" combat-ready status. Such were the trappings of a "peacetime" Army.

Spare parts would be on backorder for several months.

Repair kits would linger at their depots.

Stray tools would be snatched up by marauding mechanics.

And essential lubricants would expire unused within their cannisters.

Maintenance and monetary issues notwithstanding, the

1st Engineer Battalion had an impressive lineup of military hardware. Their parent unit, the 1st Infantry Division, was a fully "mechanized" division—meaning that their infantrymen rode into battle atop the M113 Armored Personnel Carrier (APC), and were accompanied into battle by the division's tank battalions. Thus, to keep pace with the divisional tanks and APCs, the 1st Engineer Battalion likewise had their own mechanized assets—including tactical bulldozers and armored bridge launchers.

Chuck reassured his mother that the 1st Infantry Division stood little chance of being deployed. He had heard *some* of the media's musings about the conflict in Vietnam; but it was still a "limited war." There had been no formal declaration of hostilities; no declared state of emergency; and no mass mobilization of the Army Reserve or National Guard. Moreover, Vietnam seemed to be an environment better suited for light infantry…not heavily mechanized units like the 1st Engineer Battalion.

"Well, the irony is that three months later, we had brand new equipment, and we were in the process of being deployed!"

Three weeks before the battalion began loading its equipment, however, Chuck Humphrey found himself in a new assignment. "They got another higher-ranking officer in the battalion to become the maintenance officer; and I was assigned to be a platoon leader in Charlie Company."

Captain Jay Franz, the company commander, welcomed Chuck as the new 3d Platoon Leader. All told, Chuck was excited to take command of a platoon on the eve of its first deployment. Most of his soldiers were draftees; untested in combat, but highly-motivated, and firm believers in their mission. "I had three squads of ten engineer soldiers each," said Chuck. "And I got acquainted with some real-life non-commission officers: three staff sergeants [the squad leaders] and the platoon sergeant. His name was Scroggins. He wasn't a redneck, but he sure sounded like it when he talked!"

Although the Division had just been alerted to deploy, Jay Franz had seen all the "warning signs" weeks ahead of the official announcement. "It seemed that a lot of people were being assigned to Fort Riley all of a sudden," said Jay. "I took command of Charlie Company, and with every passing day… every unit was being filled up."

It was the unmistakable sign of a unit that had been tapped for mobilization.

"Items that had been in requisition for *ages* were suddenly on hand!"

In June 1965, the 1st Infantry Division's Second Brigade arrived in Vietnam as an "advance party"—paving the way for the rest of the division. The 1st Engineer Battalion, meanwhile, readied itself to deploy with the remaining divisional units that September. From their pre-mobilization classes, however, the men of Charlie Company knew that Vietnam would be a different kind of war… and the Viet Cong would be a different kind of enemy.

The Viet Cong (VC) were guerrillas by nature. They relied on hit-and-run tactics, sabotage, spy networks, and psychological warfare. Although they were not as well-equipped or tactically proficient as the US military, these communist guerrillas were tough, resilient, and had years of combat experience on their own territory.

This would be a war with no frontlines and no rear echelon. The enemy could, theoretically, hide in and amongst the local population. Therefore, civil cooperation was of the utmost importance. American GIs had to defeat the Viet Cong on the battlefield, and simultaneously win "hearts and minds" among the local villagers in South Vietnam. Most land battles would begin as "search-and-destroy" operations, turning American patrols into hunter-killer teams. Theoretically, once the Americans had eliminated the VC from a particular area, the ARVN would step in and assume operational control.

South Vietnam itself was now divided into a series of Corps Tactical Zones (CTZs), one for each of the four ARVN corps-level headquarters. Each CTZ was an administrative and command area for tactical operations. The 1st Infantry Division, and thus the 1st Engineer Battalion, would be assigned to the III CTZ - covering the outskirts of Saigon; and stretching from the provinces of Tay Ninh in the west, to Binh Tuy in the east.

While Jay Franz readied Charlie Company for its deployment, the 1st Engineer Battalion reached its authorized strength of more than nine hundred men. Because of the anticipated terrain in Vietnam, however, the battalion would have to leave much of its heavier equipment behind. Considering the number of rivers in Vietnam (all of which overflowed during the monsoon season), MACV allowed the 1st Engineers to bring their floating bridge company. To satisfy the lighter configuration, however, "the heavier Class 60 Floating Bridge, which required a large crane to emplace the heavy steel deck bays, was replaced by the aluminum-decked M4T6 bridge." And although the armored bridge launchers would be left behind, the battalion was allowed to bring their M48-based armored bulldozers. "Special training for the 1st Engineer Battalion," at this time, "included demolitions, jungle warfare, counter-ambush techniques, Viet Cong booby traps, and the M4T6 bridge."

For Charlie Company, however, the pre-deployment phase seemed like a never-ending blur of activity. Said Chuck Humphrey: "I had been a platoon leader for only about three weeks before we deployed. But we did go on *one* short field training exercise. We went out in the field…someplace out in the training area at Fort Riley…for only two or three days in duration."

Given their time constraints, it was the best they could manage.

Still, the multi-day exercise gave them a chance to refine their tactical engineering skills under scenarios they'd likely see in Vietnam.

Then came the arduous task of loading their vehicles onto railheads. From these flatbed railcars, the 1st Engineer battalion's heavy equipment would then be loaded onto ships departing San Francisco Bay.

But loading vehicles onto a railhead was no simple task.

There had to be a well-synchronized team of ground guides, spotters, and steady-handed drivers to ensure the vehicle's proper positioning atop the railcar. A miscalculated turn of even a few inches could result in a rollover.

"It was a 24-hour operation for about a week," Chuck continued.

Teams of soldiers rotated through day-and-night shifts to ensure that every rolling stock was loaded onto the railheads.

Chuck himself had to prepare a number of Jeeps for the railway expedition.

"I remember us having to lower the windshields and tie them down, or otherwise secure them," he said. "I also remember taking the trailer that was pulled behind my Jeep." It was a small, two-wheeled utility trailer, reminiscent of those pulled by the WWII-era Willys Jeep. "It had a nice tarp covering, of course, but we put a lot of stuff in that Jeep trailer, including a foot locker," he chuckled.

Many of the soldiers likewise boarded trains westbound to the Oakland Army Terminal, where they'd meet their equipment before boarding the troop ship. Chuck Humphrey and several others, meanwhile, took a series of chartered airline flights into Oakland International Airport. "And mine departed in the middle of the night," he laughed, "out of the airport in Manhattan, Kansas."

As expected, it was a tearful goodbye for Chuck and his wife as they sat in the front seat of their '65 Plymouth Satellite. But the witty Sergeant Scroggins broke the tension by sticking his head through the passenger window, saying:

"Ma'am, I'll take good care of your husband while we're gone."

Chuck continued: "By the next morning, after we landed… we got to the troop ship [the *General Daniel I. Sultan*] and were told to go up these long ladders, and get on the ship."

And the entire transpacific journey was a "war story" unto itself.

Just like Dan Crowley's experience aboard the *Maurice Rose*, this troop carrier was a poorly-ventilated, vomit-inducing nightmare. "Five of us lieutenants leaned over the railing just before the troop ship got underway," said Chuck.

"And we jokingly said: 'Well, which one of us is going to get seasick first?'"

"Well, it was me!" he added.

"Two hours after we got out from under the Golden Gate

Bridge, the Pacific Ocean did what it was capable of doing. She rolled our ship sideways and frontwards and backwards, and I got very seasick. For the next four days, I didn't ever make it to the dining area; I actually crawled back to my room, and they [his friends] brought me some apples and soda crackers. And that's kind of what I lived on for four days."

Charlie Company landed at the port of Vung Tau in October 1965. Disembarking with the rest of the 1st Engineers, Chuck was surprised to see that their arrival was an "administrative landing." There would be no storming of the beaches á la Normandy or Iwo Jima. "They moved us onto a ship-to-shore watercraft," said Chuck—a Higgins boat of World War II vintage. "And once we got ashore...the 1st Infantry Division Band was there playing and welcoming us. That's how 'administrative' it was."

Joe Vargas, a soldier from 3d Platoon, was equally surprised. "I recall that when we arrived in Vietnam," he said, "we unloaded from the [*General Sultan*] on the side, climbing down netted ladders," and into the open bays of the Higgins landing craft waiting below. As they barreled through the surf, riding aboard their WWII-era landing craft, Joe Vargas was mentally preparing himself for a gruesome battle in the style of Omaha Beach. In fact, he was *certain* that the Viet Cong would be lining the shore, waiting to open fire.

"I knew that as soon as the front of that boat opened," he said, "I was probably going to be killed. However, to our surprise, the beach had already been secured."

Indeed, Vargas and his friends knew *nothing* about the administrative landing.

"As soon as the front of the boat opened," he continued, "we rushed toward the beach," expecting to trade fire with the Viet Cong. However, upon seeing the Division Band (and the local Vietnamese girls who had come to welcome their arrival), the young engineers sheepishly lowered their rifles and composed themselves.

"What a relief!," he sighed.

Charlie Company then climbed aboard a C-130 cargo plane

and flew into the Bien Hoa Air Base, just outside of Saigon. "Then, they transferred us all onto trucks," said Chuck Humphrey, "and we went to a staging area [at the Port of Saigon]...that's where we found all of our equipment."

From there, the 1st Engineer Battalion dispersed to the various campsites of the units they'd be supporting. Charlie Company proceeded up Highway 13 to Lai Khe, supporting the division's 3d Brigade. Battalion Headquarters, meanwhile, settled in alongside Division HQ at Di An, just north of Saigon. "At each camp, the engineer companies began constructing defenses and facilities. Initial tasks included clearing fields of fire; and building roads, portable tent floors, wood buildings, security fences, bunkers, concrete floors, and helipads."

During the weeks and months prior to the battalion's arrival, however, the "advance party" elements of the 1st Division (those who had arrived in June), had been laying the foundations for the Allied camp network...and feeling out the enemy's disposition. Although the VC were still avoiding direct engagements, the advance party troops had encountered a number of "land mines and booby traps, mostly the command-detonated directional type." Because the VC appeared to rely on demolition and sapper-style warfare, the 1st Engineers would play a critical role in neutralizing the enemy's tactics.

But developing these countermeasures was only part of the equation.

After all, Vietnam was a public relations campaign *and* a shooting war. US troops had to *show* the rural South Vietnamese that democracy and capitalism were better than anything the VC were offering. Thus, in keeping with the strategy of "Hearts & Minds," Charlie Company and the rest of the 1st Engineers received classes on Vietnamese culture, customs, and language. "It was conducted by some senior-ranking Vietnamese Army [ARVN] officers," Chuck recalled. And throughout their block of instruction, the ARVN cadre emphasized that the Americans were *guests* of a host nation, and that they should behave accordingly.

Of these introductory topics, the Vietnamese language class was perhaps the most fascinating. Learning Vietnamese was relatively easy because the language was built upon the Latin alphabet, not the symbolic characters found elsewhere in the Far East. The Latin alphabet was a gift to Vietnam from the Colonial French and their Catholic missionaries. Finding the Vietnamese script too complicated, the French devised a way to reduce the written language to a Latinized script to facilitate recording hymnal lyrics. Although it made the Vietnamese vocabulary easier to learn, it put Western learners at a disadvantage because the Latin script did not account for the differences in *tones*. Vietnamese was a tonal language, and a word's meaning could change depending on the speaker's tone and inflection.

Throughout the class, Charlie Company learned such words and phrases as:

- *Cam On* ("cahm oon"): Thank You.
- *Di Di Mau* ("dee dee mao"): Go away!
- *Dien Cai Dao* ("dee-in-kee-dao"): Crazy in the head.
- *Lai Day* ("lye dye"): Come here.
- *Lam On* ("lahm oon"): Please.
- *Nung Lai* ("noon lye"): Halt!

At the same time, the engineers began reconnoitering the area around Lai Khe. Indeed, the battalion surveyors inspected *every* road and bridge leading to the cities of Tay Ninh and Vung Tau, as well as the Phuoc Vinh Base Camp. Most of these infrastructure points were suitable for the battalion's combat vehicles, but their trafficability depended on the season. Within the III CTZ, for example, wheeled and tracked vehicles were virtually unimpeded during the dry season. During the monsoon season, however, tracked vehicles (including tanks and bulldozers) fared about a 73% "go-trafficability" rating. Meanwhile, the battalion inspected the local water wells (determining which ones could be used for Allied water points) and identified more than twenty sites for potential laterite pits. These recon teams also took soundings along the Dong Nai River, identifying the best routes to ferry heavy equipment from Saigon into Bien Hoa.

For any points of infrastructure that didn't exist, Charlie Company and their fellow engineers could simply *build* it.

Aside from their bulldozers, they had a crane, "and big, portable compressors," added Chuck Humphrey, "the trailer-mounted compressors that you could use to run jackhammers and stuff like that." Chuck also recalled having four dump trucks within his platoon of thirty-five soldiers. Of course, road construction was the most time-consuming of the battalion's missions. Depending on the needs of the local maneuver units, the roads could be as simple as dirt trails…or as complex as an improved surface with level grading and all-weather trafficability.

More often than not, however, Charlie Company had to *repair* the existing network of roads within their section of the Corps Tactical Zone. Sadly, by 1965, many of the roads beyond Saigon had fallen into various states of disrepair. Pot holes were the leading offenders, followed by craters and longitudinal cracks. "And the first mission given to my platoon," said Chuck Humphrey, "was to do some repair work on Highway 13." Running north-south, Highway 13 was the main thoroughfare in the III Corps Tactical Zone. Nicknamed "Thunder Road" by American forces, Highway 13 would become the focal point for a number of latter-day firefights between the Americans and Vietnamese Communists.

"And so, my platoon got designated as *the* Highway 13 repair crew, whereas one of the other platoons was put to work helping erect some buildings at the Lai Khe Base Camp," Chuck recalled. "And eventually, one of the other platoons was given the mission of expanding the airfield to accommodate landing a C-123 aircraft.[6]" These were massive construction projects—all the more impressive because each task fell to a single engineer platoon of thirty-five soldiers each.

Generally speaking, Charlie Company's mission was to provide "combat-related support to the infantry, whenever they asked for it." The only problem with such a broad directive, however, was that the infantry units had no formal restrictions on what they

---

[6] The C-123 was a "distant cousin" of the ever-popular C-130 Hercules. Nicknamed the "Provider," the C-123 was similar in size and shape to the C-130, and was used in conjunction with the latter until the 1980s. By design, the C-123 was a cargo/transport aircraft, but it gained notoriety during Vietnam as a purveyor of Agent Orange.

could call "support." This meant that the engineers often found themselves performing missions beyond the capacity of their manpower. Case in point: Assigning a single platoon to repair and survey Highway 13.

But Charlie Company was up to the task.

Moreover, they executed these early construction projects with record speed and quality craftsmanship.

Still, the VC saboteurs and demolitionists would never miss an opportunity to strike a "soft target" whenever they found one. During the early days of the war, they gave special attention to the French-built infrastructure. Chuck Humphrey's platoon discovered this sad fact when they were assigned to bridge-repair duty after completing their work along Highway 13. "It was a bridge right near Ben Cat," he recalled, "built by the French Army…[the structure] was equivalent to what we call a Bailey Bridge"—essentially a portable truss bridge. "It had been destroyed by the Viet Cong."

For the reconstruction, however, Chuck's platoon had to use whatever local materials they could find. "And because our base camp was on a rubber plantation," he said, "the rubber trees were our main source of timber to build and *rebuild* bridges." And although these rubber trees were sufficient to make strong bridges, the trees themselves were nearly impossible to fell. "Cutting down a rubber tree with a chainsaw is doable," said Chuck, "but it gums up the chains very fast! We had a lot of maintenance problems with that." The cutting crews often had to replace 3-4 chains for every rubber tree.

By the end of October, the 1st Infantry Division had completed the first of their "search-and-destroy" operations against the Viet Cong. General William Westmoreland, the commander of American forces in Vietnam, had granted the division "free rein to operate in a fan-shaped area opening northward from Saigon over a distance of thirty-four miles in a region ranging from grasslands to rolling forested hills."

Over the next three months, the 1st Division led numerous patrols sweeping the areas north of Saigon. Although some of these operations resulted in major firefights against the Viet Cong, the enemy guerrillas still preferred their hit-and-run tactics. Chuck Humphrey's platoon sergeant, the loveable and free-spirited Sergeant Scroggins, sadly fell victim to one such ambush along Highway 13. "He was wounded pretty bad," said Chuck, "and he was out of the unit for about six weeks, recovering at the hospital in Saigon."

During the hunter-killer patrols, however, the 1st Engineer Battalion acquired a new skill for the occasion: Clearing a helicopter landing zone (LZ).

With the onset of this new war, the helicopter had become the weapon of choice for getting troops in and out of battle—especially in areas of dense vegetation and poor ground visibility. One such battle in early November pitted a US infantry battalion against a heavily-armed VC regiment near the confluence of the Dong Nai and Song Be Rivers. "The dense jungle and close fighting made resupply and evacuation impossible" but, on the morning of November 9, 1965, engineer teams arrived via helicopter to clear additional landing zones. "With their power saws, they cut down trees, some 250 feet high and six feet in diameter," allowing the MEDEVAC choppers to pick up casualties. "The enemy, identified as a hard-core Viet Cong regiment, had been decimated, and left behind more than 400 dead."

These early engagements were critical to understanding the Viet Cong's tactics and techniques, *but* American forces had yet to meet the North Vietnamese Army (NVA) in combat. However, on the morning of November 14, 1965, heliborne troops from the 1st Battalion, 7th Cavalry squared off against two NVA regiments at the Battle of Ia Drang. It was the first major battle between US and North Vietnamese regulars. Although outnumbered four-to-one, the Americans won a decisive victory, claiming the lives of more than 1,200 enemy soldiers.

Still, the After-Battle Reports confirmed that the NVA were no joke.

They were well-trained, well-disciplined, and possessed a sort of "suicidal fanaticism." Despite taking heavy casualties, they had repeatedly attacked the American lines with unflinching aggression. The NVA, however, were likewise impressed by the Americans—especially in their coordinated use of artillery and helicopters. Almost immediately, the NVA started formulating counter-tactics to defeat (or at least neutralize) the Americans' helicopter advantage. The most effective way, it seemed, was to draw American ground forces underneath the jungle canopy (thus concealing them from any close air support), or continue setting up ambushes and obstacles along the main arteries in South Vietnam.

From mid-November until late December, the 1st Engineers patrolled a handful of search-and-destroy operations between Highway 13 and the Michelin Rubber Plantation. During these sweeps, the deadliest firefights erupted from "chance encounters [and] enemy ambushes on returning convoys." And with every search-and-destroy mission, the 1st Engineers uncovered more and more enemy caches. For each patrol, Charlie Company provided a "composite platoon and demolition teams," to destroy all VC equipment captured by the infantry.

During their inaugural year of combat, Charlie Company's casualties were light…but they were painful losses nonetheless. As Jay Franz solemnly remembered: "I had to identify the remains of my men who were killed on missions." It was *the* most agonizing job for any commander in combat. "I could send out a handful of guys in the morning to go along with the infantry," he said, "and you hoped that the same handful came back. But too often they didn't."

To make matters worse, Jay also witnessed the improbable demise of Lieutenant Colonel Gordon J. Lippman, the executive officer (XO) of 3d Brigade, 1st Infantry Division. As the Brigade XO, Lippman had been the primary liaison in coordinating engineer support for 3d Brigade's operations.

On the night of December 11, 1965, Lippman invited Jay to dinner at Brigade Headquarters. "Well, the brigade commander was living in what had been the [rubber] plantation overseer's house," said Jay. "And most of the brigade staff lived there. It was a pretty good-sized house." Most of its antebellum beauty, however, had been robbed by the sudden onset of sandbags and barbed wire perimeters. "So, I went down there," Jay continued, "and it was the first time on my tour that I got to sit at an actual table with real silverware!"

The meal was posh by Army standards, and the officers were having such a good time that they hardly noticed the sniper fire in the distance. "We had sniper fire in the perimeter just about every night," said Jay—so they had grown fairly accustomed to it. By now, most GIs considered the sniper fire more of a nuisance than an actual threat.

But tonight would be different.

"After the meal, I remember the XO getting up and putting on his web gear, which included the .45 holster." The term "web gear" referred to the tactical belt and load-bearing suspenders that soldiers wore in the field. This accoutrement carried a soldiers' canteens, grenades, side arm, and extra ammunition.

"I'm going out to see how the boys are doing," said Lippman.

And by "boys," he meant the MPs guarding the Colonel's quarters.

"Probably less than a minute after he walked out," Jay recalled, "We heard somebody outside in the dark yelling that the colonel's been hit."

Jay ran out into the darkness at a hurried pace.

"I was the first one up there," he said—and he skidded to a halt along one of the trails leading to the MP's guard station. "Colonel Lippman was stretched out on one of the trails…and he'd been shot in the back."

Lippman was alive and still conscious, but he knew he'd been seriously wounded.

At first, Jay saw a gleam of hope because: "There was a MEDEVAC helicopter on the pad…and there normally wasn't. Typically, they all went back to Saigon at night."

But, for whatever reason, this particular chopper had stayed behind.

"So, we considered that a really good omen."

As the flight crew began cranking up their helicopter, getting ready for the emergency flight into Saigon, Colonel Lippman lay stretched out atop a conference table in one of the rooms at Brigade Headquarters.

And Lippman knew the end was near.

"He was talking to the brigade commander," Jay continued, "and it sounded like a speech you would hear in a movie…about what an honor it had been to serve with him." Indeed, Lippman's words sounded like the departure speech that a dying character would give his comrades before he expired. "I can't remember all the words, but it was moving to save the least." And as the MEDEVAC crew hoisted Lippman out of the room, his final words to Jay and the other officers were:

"I won't be back."

Still, Jay and the others were optimistic that he would make a full recovery. After all, Lippman was now a *three-war veteran*. He had fought in the Battle of the Bulge; and had earned the Distinguished Service Cross for his actions at the Hantan River in Korea. Surely, a man of that caliber and pedigree wouldn't come so far…only to be felled by a random sniper in Vietnam.

Sadly, Gordon Joseph Lippman didn't survive.

"And we were devastated to learn the next morning that he had died. That was a real loss."

On December 30, Dan Crowley stepped off the bus at Battalion Headquarters in Di An. He was one of twelve individual "replacements" who would be joining the 1st Engineer line companies.

Being a replacement, however, was no enviable position.

Very often, their job was to *replace* an individual soldier who had been killed in action. As expected, these replacements were commonly seen as outsiders—green soldiers who had to prove their worth, and try to make friends among the battle-hardened veterans in their company. Too often, replacements came and went so quickly (killed or wounded) that their platoon mates avoided getting to know them.

Luckily for Dan, it was still early in the war, and the battalion's casualties had been comparatively light. "I was assigned to Company C—Charlie Company—whose callsign was 'Die Hard

Charlie.'" For Dan Crowley, this callsign was both captivating *and* ironically amusing. Per NATO's Phonetic Alphabet, all lettered "C" companies were designated "Charlie." But at the same time, American forces in Vietnam also referred to the Viet Cong as "Charlie."

As it turned out, Charlie Company had just convoyed into Di An from Lai Khe a few days earlier. Still supporting 3d Brigade, the company was preparing for a new mission:

Operation CRIMP.

"The 25th Infantry Division was on their way from Hawaii," said Dan. And this incoming division would occupy the Cu Chi Base Camp, northwest of Saigon. Charlie Company's mission (as part of 3d Brigade's effort) was to eliminate any Viet Cong forces near Cu Chi. Ideally, this would allow the 25th Infantry Division to establish its foothold without interference from enemy fire.

Simultaneously, Operation CRIMP was targeting a suspected Viet Cong headquarters, purported to exist within a series of underground tunnels near Cu Chi. South Vietnamese agents, VC prisoners, and aerial reconnaissance photos indicated that this subterranean network lie somewhere north of the Cu Chi Base Camp.

Dan Crowley, meanwhile, found himself assigned to 3d Squad; 1st Platoon. "My squad leader was Staff Sergeant Ron Riley…a big man with a red handlebar mustache and a 'take-no-bullshit' look." Indeed, Riley was one of the most impressive and intimidating soldiers that Crowley had ever met.[7] "He tells me that I'm to be one of his two Demo Men," Crowley continued. "I'm replacing one who was KIA…but I will get an extra $55 per month for Demo Pay." Dan was none too happy to be replacing a KIA soldier, but he was grateful for the opportunity to do what

---

[7] Dan Crowley recalled that, after the war, he and Ron Riley became close friends. For several years, they would tag up at the 1st Engineer Battalion Reunions. It was during these reunions that Crowley learned much more about his former squad leader. For example, Riley had deployed to Vietnam earlier in 1965 as part of the 1st Infantry Division's "advance party," and returned for a second tour of duty in 1970. Riley also became an Army lifer, retiring at the rank of Sergeant Major. Sadly, Ron passed away in 2019.

he did best: demolitions and pyrotechnics.

"I was issued an M-79 Grenade launcher," he said, "along with a 45-caliber pistol and canvas satchel bag…for carrying explosives." His satchel—the so-called "Demo Bag"—was *the* essential tool kit for a demolition man in Vietnam. It contained eleven blocks of TNT (five one-pound blocks; six half-pound blocks); twin blocks of C4 explosive (two pounds each); time fuse; detonator cord; blasting caps; blasting machine; trip wire; and fuse lighters.

"Christmas at last!" he joked.

Dan spent New Years Eve 1965 inventorying his new demolition gear…and fondly remembering his days as a young firecracker tycoon.

Of all the items in his carry-on arsenal, however, Dan Crowley had a special affection for the M-79. Nicknamed "Thumper" and "The Bloop Tube," the M-79 was capable of firing 40mm grenades and specially-modified cannister rounds that discharged ball bearings similar to a shotgun blast. In fact, Dan recalled that the M-79 resembled a single-barrel, sawed-off shotgun. "It was accurate to 300 yards without much kick," meaning that the M-79 didn't have the jolting recoil found in other handheld weapons.

But as Crowley would soon discover, the life of a combat demolitionist left "no room for error."

You were either skilled…or dead. There was nothing in between.

"A demolition expert has the responsibility of felling trees to clear the way for air strips and helipads. He must be equipped to detect and destroy enemy mines and booby traps"—the latter of which was a perennial favorite for the Viet Cong. Booby traps came in all shapes and sizes; but the VC had a special talent for rigging explosives to the most seemingly-innocuous items. For example: "an empty C-ration can, which an unsuspecting GI might kick out of his path." A good rule of thumb, therefore, was to leave any manmade object where it lay—until a demolition team could evaluate it for potential death traps.

Operation CRIMP began on January 8, 1966. It would be Dan Crowley's first time in combat. Charlie Company convoyed along

Highway 1 from Di An towards Cu Chi—linking up with an ARVN Ranger unit at their outpost in Trung Lap.

"From afar, the place looked eerie," said Dan.

Charlie Company's orders were simple: Tie into the ARVN Rangers' perimeter and dig in. On his first night in Trung Lap, Crowley discovered that the nighttime sounds were nearly as spooky as the camp itself. While on guard duty that night, he was frequently startled by some strange chirping and hissing sounds.

"I was thinking that the VC were messing with me."

He woke Sergeant Riley to inform him of the strange noises. But the ginger squad leader simply rolled over and told him that the "noises" were just the local lizards enjoying their nighttime scavenger hunts.

But nocturnal reptilians seemed to be the least of Dan's worries.

"My uniform was worn out." Like many incoming soldiers, his green fatigues bore the white name tapes over the left breast pocket; and the black-and-gold "US Army" tapes above the right breast pocket. While these accoutrements looked nice in garrison, they weren't easily camouflaged in the jungle. In short, he needed jungle fatigues with subdued lettering…and a sturdy pair of jungle boots. He had been in Vietnam for only a few days, and his stateside-issued boots were already feeling the strain of the jungle environment. And as luck would have it, his unit's supply room had no jungle fatigues or boots on hand. Thus, Dan had little choice but to carry on with his ill-conceived uniform and deteriorating footwear.

"I reported to Sergeant Boldin, the 1st Platoon demo leader," said Dan. "I told him about my boots, and he tells me we're going on a search-and-destroy mission with the infantry, and that I should draw another block of C4."

Dan was confused.

Another block of C4? Why?

No amount of C4 could fix his uniform issues.

But Dan was about to learn that C4 was a currency unto itself. "It was a valuable commodity," he said—"used to heat C-Rations and coffee. All you had to do is take a piece the size of a dime, place it on the ground, and touch a match to it. It burns a hot blue flame." And the infantrymen were willing to trade nearly *anything* for a spot of C4—even a fresh pair of jungle fatigues and matching boots. Now he understood why Boldin had made him bring the additional C4.

"We were the 'Candy Man' to these grunts," Dan chuckled.

Aside from the new uniform items, Dan also bartered his C4 for extra food items, cigarettes, and M-79 cannister rounds. It was amazing what a few slices of C4 could command on the GI trade market.

CRIMP was the first division-sized operation of the war, and the largest Allied military action in Vietnam to that point. Fighting alongside the Americans was the 1st Battalion, Royal Australian Regiment (1RAR). Nominally a motorized rifle battalion, 1RAR had deployed to Vietnam the previous March as part of the Allied effort against the Viet Cong. "Those Aussies were real bad asses," said Joe Vargas. "You could say that they 'took no prisoners' and they were not afraid of anything." In the coming months, Charlie Company engineers would go on a number of joint patrols with their Australian counterparts.

On the first day of Operation CRIMP, Dan Crowley hopped aboard a UH-1 "Huey" helicopter with the rest of his demo team. They would be inserted onto an LZ a few miles out, joining 2-28 Infantry in their massive search-and-destroy operation. "It was my first time in a Huey," Dan recalled. "What a ride! We were told by the crew chief that snipers had shot at choppers…and we were to un-ass as soon as the skids touched down!"

But today, the pilot wasn't taking any chances.

The helicopter crew pushed Dan and his comrades out of the aircraft while it was still hovering three feet off the ground. After the crew tossed the last engineer from their bay, the helicopter hastily fled the scene.

"So much for a touchdown," said Dan.

"We hauled ass to the tree line," he continued—and he breathed a sigh of relief when he realized they hadn't landed under enemy fire. "We were now with the headquarters unit of the 2d of the 28th Infantry."

For the next six days, his demolition team rotated amongst the infantry companies—providing demo support or acting as provisional foot soldiers. Sometimes, the demo team would have prior notice before going on a patrol...and sometimes not. Dan characterized these cold calls for a pick-up mission as a matter of "*Grab your gear and jump through your ass.*" Indeed, they would grab their demolition kits and run to the nearest helipad, very often not knowing their destination until they were airborne.

After touching down at any given LZ, Dan and his fellow engineers simply walked alongside the infantry.

"Imagine a herd of buffalo looking for the enemy," he said.

Whenever they stopped for the night, they would dig two-man foxholes. Each occupant would then rotate through a guard schedule: One man would sleep while the other stood guard, rotating shifts every two hours or so.

For Dan Crowley, Operation CRIMP was light on enemy contact, but very heavy on equipment finds. "Lucky for us," he said, "we had very little contact with the VC...we did, however, find all kinds of stuff to detonate—dud mortar rounds, bunkers, booby traps, and aerial bombs. But no 'Charlie.'"

Admittedly, the bunkers fascinated him the most. These were the terminals of the vast underground cave complex that they had read about in the Allied Intelligence reports.

In fact, these were the notorious *Cu Chi Tunnels*.

"The tunnel system went down three, four, or more levels to the water table, and linked the local villages and hamlets. While some of the tunnels were newly dug, others were so old that moss grew on the walls." At various points along the tunnel network, the VC had built a tiered system of underground rooms. These included medical clinics, storage areas, barracks, command centers, and galleys.

All told, it was an impressive feat of engineering.

These subterranean tunnels showed that the NVA and Viet Cong understood the principles of structural engineering and soil dynamics.

As the VC fled from the Allied dragnet, Charlie Company

began to explore their vast underground network. "That was our first experience of trying to investigate these tunnels, and find out how extensive they were," said Chuck Humphrey. "Three men in my platoon volunteered to do it. At first, they were called 'tunnel explorers.' But as the war went on, they became known as 'tunnel rats.'"

Chuck discovered that many of these tunnels dated back to the 1940s, built by the Viet Minh during their war against the French. "The tunnels were abandoned after the French left the country," said Chuck. "But as our involvement in South Vietnam ramped up, the Viet Cong began reusing those same tunnels."

To secure these enemy caverns, Chuck Humphrey's tunnel rats had the dubious honor of blasting CS gas into the complex. CS was a non-lethal, specially-modified tear gas designed to restrict one's breathing and agitate the mucus membranes. "Our battalion commander showed up with these 50-pound bags of what turned out to be the granular form of CS gas," said Chuck. "The game plan was that we would put these bags of CS granules far enough into the tunnels, wrap them in detonating cord, and then get safely out of the tunnel before setting off the explosion. And the logic was that the CS gas would penetrate these tunnels and make them uninhabitable."

In other words, they were trying to smoke out the VC.

"Well, it didn't work," he admitted.

Elsewhere along the Cu Chi network, other Allied units were trying similar tactics. They set up high-powered air pumps, forcing high-pressure concentrations of smoke and tear gas into the tunnels. These units, too, were hoping to drive out the Viet Cong and expose other entrances. But the VC entrance covers were simply too heavy to be shaken by air pressure.

By the second day of Operation CRIMP, Dan Crowley and his demolition team had covered a lot of ground. Aside from the occasional (and wildly inaccurate) VC sniper fire, the engineers continued their patrol unmolested. As Dan recalled, their second-day yields included: "Punji stakes[8], spider holes, bunkers, trench lines, bags of rice—but no VC."

On the third day, however, Dan Crowley had his first encounter with a friendly KIA. For most of the morning and afternoon, their Day 3 patrol had been a repeat of the previous two days—"a grenade here, a booby trap there, and a dud mortar round." Late in the afternoon, however, the patrol leader decided to halt and dig in for the night. But as the Americans were setting up their bivouac, a lingering VC sniper took aim at the halted patrol.

Dan jumped with a startle as the enemy shot rang out into the twilight.

"A minute later," he said, "two GIs are carrying and dragging a third GI." Their wounded comrade had been hit center-chest. He died moments later.

Witnessing a friendly KIA was devastating enough, but nothing could prepare Dan for the shock of learning that the 2-28 Battalion commander, Lieutenant Colonel George Eyster, had also been killed. Eyster was a 1945 West Point graduate and, like many of his classmates, he had served on the frontlines in Korea. Now in the throes of Operation CRIMP, Eyster had become another victim of VC sniper fire.

At first, however, his gunshot wound appeared to be minor.

And within the first 48 hours, it seemed as though Eyster would make a full recovery. However, he suffered the abrupt onset of a pulmonary embolism and passed away on January 14, 1966.

The aftermath of Eyster's death, however, highlighted the moral support that Americans were still giving to their military at this stage in the war. Indeed, Eyster's widow received more than 2,000 letters from strangers offering their condolences. After reading her husband's obituary, she felt compelled to thank the war correspondent who had published it. She wrote: "You gave his children a legacy that no one else could have by writing in

---

[8] Punji stakes (also called "Punji sticks") would become a persistent nuisance throughout the war. They were simple stakes, often made from bamboo. The name itself came from the colonial forces in latter-day British India, describing the wooden stakes they encountered during the Anglo-Burmese border conflicts. In Vietnam, Punji stakes were often placed within camouflaged holes, or along the tactical routes travelled by Allied forces. They were sharp enough to puncture most kinds of tactical footwear; and the VC would occasionally lace the sharp ends with poison or fecal matter.

such a manner that his courage and heroism will live with them and be an inspiration to them forever."

Meanwhile, the other platoons in Charlie Company had been fighting the Viet Cong near the edges of Trung Lap. While Dan Crowley's demo team was patrolling with 2-28 Infantry, one of the Division's helicopter pilots reported that some VC operatives were laying mines along the road to Highway 1. As Jay Franz recalled: "I had visited HQ on some other matter when one of the officers—possibly the Brigade S3—passed that report to me,"—asking Jay to send a platoon down the road for a hasty reconnaissance.

"So, I did send one platoon, but I went with it," said Jay.

"And in retrospect, it was obviously a trap."

First Platoon (minus the demo team attached to 2-28 Infantry) left the perimeter with only *one* Jeep and three tactical dump trucks as a means to complete their route reconnaissance. The 5-ton dump truck was standard inventory among the heavy engineering units, but it offered little in the way of maneuverability or protection. Still, it was the only mobile platform that Jay had available at the time.

A few minutes after leaving the gate, 1st Platoon deployed their minesweepers along the road. The non-minesweeping troops, meanwhile, settled into their hasty defensive positions, providing overwatch as the sweep began.

"Suddenly, we came under fire," said Jay.

As the VC opened fire from all directions, he scrambled for the radio, calling for immediate air support.

"And by golly, we had gunships in the area," he said.

A flight of UH-1s answered the call, thundering overhead and strafing the designated enemy position. The sudden appearance of Huey gunships was enough to force the VC's heads down; but one pilot's adrenaline was running so high that he overran the target area, accidentally strafing PFC Henrick, a 1st Platoon soldier. Fortunately, Henrick survived the incident—albeit severely wounded in his buttocks.

Near-fratricide incidents notwithstanding, Jay was pleased that the close air support had arrived so quickly.

"They really hosed down the area," he said.

Jay also remarked that, if not for the underground tunnel network, any onlookers to the strafing would have said: "Nobody could survive that."

But today, the Viet Cong had survived.

Some VC had been caught by the Huey gunfire, but more had simply jumped down into their tunnels and spider holes, emerging to resume their small-arms fire after the gunships departed.

The soldiers resumed their minesweeping—"with no mines found, of course," Jay added—and they were about to leave when the enemy fire erupted again. "By the time they sprung that ambush again," Jay continued, "every bit of helicopter air power was reloading, rearming, or refueling…and they were not available. It wasn't clear until later that the VC were just bugging out, running into their caves. That cave business was unknown to us at the time. We learned the hard way."

By this point during the firefight, however, there had been no fatalities or serious injuries. Staff Sergeant Ron Riley, however, radioed that one of his men had suffered a serious hand injury—a finger had been shot off. This unlucky soldier seemed to be going into shock, which could very easily lead to death. "It seemed important to evacuate that man, if possible," said Jay. Determined to save that soldier from the jaws of a well-planned ambush, Jay drove to his location, and waited the few seconds while Riley hoisted the injured man into the back of Jay Franz's Jeep. Riley jumped into the passenger seat himself, and the adrenaline-addled trio drove out of the kill zone the same way they had come in. "Despite the intense firing," said Jay, "no rounds hit us or the Jeep."

It was nothing short of a miracle. But Jay's luck was about to run out.

"After getting the man to the Aid Station," he continued, "I gave Riley the option of staying there at HQ to help organize some relief." But Ron Riley, never one for "opting out," declined the offer, and jumped back into the Jeep. "We drove together back into the ambush area."

Finding the platoon leader, Lieutenant Joe Coppolo, Jay

Franz ordered the convoy's evacuation. "We determined it was feasible to leave the area as we had come—in our vehicles," said Jay. "Jumping onto those dump trucks was going to be risky, but staying there seemed riskier."

Suddenly, Sergeant First Class Carroll (the platoon sergeant), rumbled into view. The old man stood atop the running board of 3d Squad's dump truck—"shouting for his men to jump aboard the slowly-moving vehicles," Jay recalled. "I had just started back toward my Jeep when I heard an explosion and felt the shock wave." Jay spun around just in time to see the dump truck—"and what turned out to be Carroll's body"—flying through the air. *The truck had just rolled over a VC mine*. Surprisingly, the driver and a few other men survived with less-critical wounds. "Very soon after that explosion," Jay continued, "I was injured by a bullet that hit my left arm and destroyed my shoulder."

Due to the severity of his wounds, Jay Franz was medically evacuated from Vietnam. "I spent about two weeks at the 93d Evacuation Hospital, basically on a cot…then one night in the 3d Field Hospital in Saigon," en route to San Francisco for the remainder of his convalescence.

As Jay Franz was being evacuated, Larry Blair was en route to Vietnam; and he would ultimately replace Franz as the company commander. Initially, however, Larry had been slated to assume command of Alpha Company after its commander had been killed by an enemy mine. "Of course, by the time I got there," said Larry, "they had already filled that slot;" another captain had been placed in command of Alpha Company. "But then C Company's commander got wounded and evacuated," referring to Jay Franz.

"I wanted to take over Charlie Company," he admitted, but the new battalion commander, Lieutenant Colonel Joe Kiernan, had other ideas. Since Jay's evacuation, Captain Robert Zilenski had assumed command of Charlie Company, and Kiernan wanted to give him more rated time in command. Thus, Larry Blair lingered in the Battalion Headquarters before taking command of Charlie Company later that summer.

Chuck Humphrey, meanwhile, had been trying to buoy the morale in 3d Platoon. In the wake of 1st Platoon's ill-fated minesweeping, "my platoon went out there the next morning to help clean up." Picking up the pieces was a horrific experience unto itself—especially when they recovered Sergeant Carroll's body.

"The explosion had cut his body into four pieces," said Chuck.

As it turned out, the "landmine" that killed him was a primitive, subterranean, improvised explosive device (IED). In a grim foreshadowing of what American troops would see during the Iraq War, this IED was: "an artillery shell that the VC had set up with buried wire, so they could detonate it whenever they wanted." Today, these are known as "command-detonated IEDs."

The casualty cleanup had been horrific enough; but 3d Platoon's bad day was about to get worse. "Later in the evening that same day," Chuck continued, "I heard something that sounded like a mortar tube…*ploomp!*"—the unmistakable sound of a launching projectile. From the outdoor acoustics, Chuck could tell that this enemy mortar was far beyond the perimeter.

But relative distance wasn't the problem.

The mortar was still close enough to hear it.

"So, I wasn't surprised when it landed right in the area where we were bivouacked," he said. "Apparently, the enemy had done enough estimating and calibrating to hit our unit, because that first round landed right in the middle of some trucks."

Fuel trucks no less.

"They went up in flames."

The mortar attack continued into the pre-dawn hours. "It got so intense," said Chuck, "that a nearby artillery unit moved one of their towed, 105mm howitzers right alongside us on the perimeter." Levelling its gun tube, the artillerymen fired their howitzer directly into the suspected position of the enemy mortar team.

Chuck was impressed.

No calculations; no adjustable trajectory; just levelling a cannon for direct-fire use.

"It was effective," he said, "because the mortaring stopped."

The following day, Chuck's men were selected for another repair mission—this time for some roadwork along the main trail leading out of Trung Lap. "And, for whatever reason," he said,

"I decided to have my Jeep go first and lead the dump trucks." Directing his driver to the head of the column, Chuck settled into the passenger seat, checking his map as he led the convoy beyond the gate.

"Well, we hadn't gotten a quarter-mile out of the base camp," he said, "when I heard this explosion."

Something in the road had detonated behind him.

"I turned around…and the hood of the first dump truck was flying through the air!" That truck had been the second vehicle in the convoy, only a few yards behind Chuck Humphrey's Jeep. The truck was totaled, but its driver emerged with only minor injuries.

Still, Chuck was confused.

Obviously, the truck had hit an IED.

But this truck had been travelling *behind* him in the convoy.

Logically, that IED should have exploded underneath his own Jeep, not the follow-on truck. What, then, had accounted for the delay?

Upon closer inspection, Chuck realized that this roadway bomb had been a pressure-activated IED. Just as before, the Viet Cong had rigged an artillery shell. But this time, "they had set it up with a little detonation device wrapped in banana leaves" and set the device so that it lay right in the path of the US convoys. Thus, whenever a passing vehicle encountered the device, the vehicle's own weight would trigger its detonation.

"But apparently my Jeep wheels weren't heavy enough to set it off," he said wryly.

"So, I'm still here."

Chuck conceded, however, that the IED had been an unsettling "near-miss" with death. "I could have been killed that day."

The next morning, Dan Crowley and his demo team returned to Trung Lap. "We had been gone for six days," he recalled. Yet, his comrades at Trung Lap were acting much differently now. "Things had changed in our perimeter."

He soon found out why.

Dan was perplexed by the news that greeted him: Sergeant Carroll's death; Captain Franz's evacuation; multiple WIAs within the company; mortar attacks; demolished fuel carriers; and the growing nuisance of IEDs. It all seemed surreal.

The disabled dump truck, however, was one of the first things to grab his attention. "The front end was missing…the headache board over the cab was bent up at an odd angle,"—a victim of the Viet Cong's IEDs. "Remember, this is a 5-ton dump truck!" Dan could hardly believe that an improvised landmine would cause *this* much damage to a 5-ton vehicle.

Crowley then went searching for his duffle bag and foot locker. Hopefully, his personal effects hadn't been destroyed by the mortar attack. "The duffle was ok," he sighed, "but the foot locker was seriously mangled."

Indeed, the poor foot locker had been hit by flying shrapnel. "The inside was a rat's nest of shredded clothes and a punctured can of shaving cream." Whether being pilfered by his comrades in Oakland, or riddled by VC mortar fire, Dan Crowley's personal gear never seemed to catch a break.

But Operation CRIMP had ended. By all accounts, it had been a resounding tactical victory for Allied forces in Vietnam. The US-Australian task force had confirmed 128 Viet Cong killed, with an additional 190 probable kills and 92 captured. Allied losses were 22 killed-in-action (14 American; 8 Australian).

The biggest yield of the operation, however, had been the discovery of the Cu Chi Tunnels. Within the catacombs of that VC network, American forces had uncovered thousands of critical documents. On one occasion, an Australian tunnel rat delivered a satchel containing the "master file of detailed information on all the Viet Cong in the region." At the higher levels of MACV, Operation CRIMP was being heralded as a strategic intelligence victory.

Sadly, that victory would soon fade. Although MACV had now declared the area "secure," the Viet Cong soon re-occupied Cu Chi, and used their previously-vacated tunnels as a staging area for the 1968 Tet Offensive.

But for now, the only concern among the men in Charlie Company was getting some rest before their next operation. They had emerged from six days of continuous combat. The VC had fled…and they were still on the run.

CHAPTER 4
BLOOD ALLEY

Fresh from the throes of Operation CRIMP, Dan Crowley received a two-day pass to Saigon. These two-, three-, and four-day passes were the most common form of giving soldiers paid time off. In times of war, a soldier on pass could travel nearly anywhere within the "secured" sectors of an operational theater. During World War II, for example, soldiers could take passes to cities like London, Rome, Tunis, and Algiers. In South Vietnam, however, Saigon was the preferred destination.

But even in Saigon, a GI had to be on his guard. Viet Cong spies lurked amongst the local hotels, bars, markets, and bordellos. Some VC were even moonlighting as street peddlers. Thus, before every jaunt into Saigon, US servicemen had to be briefed on MACV's security considerations. GIs were advised not to discuss any military operations in public, even with other US personnel. Fraternization with local Vietnamese women, while not prohibited, was strongly discouraged. After all, a romantic liaison could unwittingly lead to a Communist entanglement. Still, that didn't dissuade many a GI from seeking company among Saigon's working girls.

"We were put in an old French Hotel," said Dan. He and the rest of Charlie Company now had two days to decompress in the lap of luxury.

Not surprisingly, most of them went to the local bars and bazaars.

But for Dan Crowley, instead of getting a beer, he was determined to buy a jungle hammock. "In South Florida, I had one and I used it often when camping." Now, during his long patrols in the jungle, a hammock would certainly come in handy.

Perusing the Saigon marketplace, however, Crowley was astounded. "My God," he exclaimed, "some of the stores had more supplies than our own supply room!" And the variety of goods was even more astounding—"Jeeps, guns, meds, booze"—and plenty of cheap thrills from the local *femme fatales*. "I nearly went broke," he confessed.

At the conclusion of their pass, Charlie Company returned to their base camp at Lai Khe. The US combat mission was barely a year old…yet Lai Khe was already showing the signs of wartime blight. "Our tents had rot and holes"—Dan continued—caused by shrapnel, VC mortars, and accidental discharges from sleepy-eyed soldiers.

These punctured tents were tolerable, but the local wildlife had become *insufferable*.

Monkeys were the worst offenders. Dan often referred to them as "nasty little bastards." True to their primitive instincts, they freely urinated throughout the camp and tossed their own feces at any passersby. "We had a female dog on the base camp,"—a local mutt that one of the soldiers had taken in—"and they [the monkeys] tried to screw it all the time. Then when you tried to chase them off, they'd sling a turd at you…or climb a tree and piss on you. And they'd steal anything they could get their hands on…including hand grenades, ammo, mirrors, anything with a shine to it."

A few fed-up soldiers began targeting the aggressive simians with any spare ammunition they could muster. Others, like Joe Vargas, however, tried their luck at *taming* the wild primates. "I had a pet monkey," he beamed. "We had them running all over the base area…and I captured a little one and kept him close." His trusty companionship, however, didn't last. Naturally, the chain of command didn't approve of soldiers keeping exotic pets while on deployment.

By late January, Crowley's demo team had been reassigned

to Highway 13, conducting "repairs and sweeps for mines" interspersed with the normal variety of search-and-destroy missions. During one of these patrols, Crowley had an unexpected front-row seat to a B-52 airstrike. That day, Dan's patrol had been sweeping through the jungles in and around Loc Ninh. "The area had rolling hills, and off to the north, it opened up with a wide view of the jungle." Now on a hilltop, their panoramic view was interrupted by the sudden appearance of an incoming Huey. When it landed, the 1st Infantry Division commander, emerged with his General Staff.

"They un-ass the Huey and consult a map," Dan continued.

"One officer points off to the north," and moments later, a flight of B-52s appeared at 30,000 feet. "There is a low rumble. The Earth starts to move. Crunch, Boom…rolling smoke with debris and flames. Then, in a flash, a big piece of jungle is eradicated."

Amazed by the B-52s' handiwork, Crowley wanted to explore the impact zone, but his patrol had to continue along their prescribed mission route. Months later, however, he was able to visit the impact zone, whereupon he noticed that everything was "bent and busted…huge crater, trees uprooted…it was awesome. I was impressed and would have hated being on the receiving end. Nothing could have survived"—not even a VC tunnel; the crater was that deep.

As January turned to February, Charlie Company readied itself for Operation ROLLING STONE. The objective was to "construct new roads and repair existing parts of 'Route Orange' between Route 13 at Ben Cat, east to Route 16, so that it could be used to speed supplies across an all-weather main supply route to Phuoc Vinh."

Ground reconnaissance had determined that the engineers could build a 12-mile road across a stretch of terrain encompassing dense jungle, plantation groves, and rice paddies near Ben Cat. This terrain, although dense and biodiverse, was level enough to accommodate a single road without any bridges or culverts. For Charlie Company, this was a welcomed relief. Bridges and culverts were tedious to build, and often drew attention from the Viet Cong. Moreover, Operation ROLLING STONE would take Charlie Company out of the search-and-destroy missions, and return them to their core mission of combat construction.

Thus, on February 7, 1966, Charlie Company climbed aboard their convoy into Ben Cat. "The company leaves Lai Khe heading south on Route 13," said Dan. Along the way, Dan's truck passed an ARVN compound where, from beyond the perimeter, he was stunned to see a "collection of about 40 bomblets in a pile—supposedly all duds. I couldn't figure out who would risk life and limb to move something so potentially dangerous instead of just blowing them in place. It was a mystery to me, but as I said, I was only a PFC." Still, this questionable ordnance disposal highlighted more differences between how the US and ARVN conducted their daily operations.

Arriving at Ben Cat, Dan and his team began sweeping the area for landmines. "We found lots of stuff, mostly old duds of different sizes and shapes," he recalled. "We blew them in place." The following morning, however, Dan was shocked to hear that the company cook was dead.

But how?

Cooks were never supposed to leave the wire.

Sadly, though, the company cook had left the perimeter with ulterior motives.

"He went AWOL in the night," said Dan, "and his body was found on the road to Ben Cat." Still, Dan was befuddled.

"Why?" he asked.

"Where are you going to go AWOL in Vietnam?"

No one knew why the errant cook had gone beyond the wire…or where he had intended to go. But the VC had obviously killed him during the night.

On the afternoon of February 22, 1966, while Dan Crowley, Sergeant Boldin, and Specialist Dan Kingbird (the other man within their demolition team) were clearing stumps for the road construction: "We notice that our infantry security team was tearing down their hooches [tactical sunshades]. They were being helicoptered into a nearby battle and, within a matter of minutes, they were gone."

Dan was worried to see the infantry leave in such haste. But it was the nature of the business.

The infantry had to be "on-call" for fellow units engaged in heavy fighting.

But the engineers still needed an infantry unit to provide security while they continued building the road. "We're on our own, but help is on the way."

And that "help" arrived in the form of 1RAR.

"An Australian long-range patrol was coming into our area." Crowley's platoon leader, Lieutenant Joe Coppolo, tapped him and a few others to set up a rendezvous point for the incoming Australians. "About an hour later," said Dan, "they show themselves." And their first question to the American outpost was: *"Got any beer, mate?"*

That night, Charlie Company dug foxholes and set a hasty perimeter, lined with barbed wire and trip flares. "Two Aussies held the northeast corner where the road met the rubber trees," Dan recalled. "One had an M-60 and the other had an M-16 with a starlight scope." Dan Crowley, along with fellow GIs Dave Smith and Dan Kingbird, made up a three-man team covering a 40-foot field of fire, adjacent to the two Aussie gunners.

Just as before, Charlie Company's foxhole teams rotated through a sleep-guard schedule. But while Dan Crowley was enjoying his slumber, Dave Smith jolted him awake, claiming that he heard voices speaking Vietnamese.

"I heard it too," said Dan. "Oh crap!"

At nighttime, the perimeter became a Free Fire Zone—meaning that any soldier could fire at will if he suspected enemy movement beyond the wire.

When Dan settled into the foxhole, however, all hell broke loose.

"The Aussies' M-60 goes *rat-a-tat-tat*. Long, long bursts!" The M-60 stopped just momentarily before the Aussie gunner fired off another deafening burst.

"What the f*ck?!" Crowley exclaimed. "Talk about pucker factor!"

He couldn't see what the Australians were firing at, but he

had little choice than to stay focused on his own sector, lest this firefight be the opening rounds of a broader VC probe. "All this is going down really fast," Dan continued. "Then, from down the road, a .50 caliber opens up."

At first, Dan and his comrades were elated by the sound. The .50 cal was an American weapon. So, whoever was firing it, they had to be joining the firefight to disperse the VC. But Dan's excitement quickly turned to terror when the tree branches above his foxhole exploded from the gunfire.

The .50 cal was shooting in his direction.

"Oh sh*t!" he cried.

The next burst of .50 cal bullets came in closer, skimming across the top of the foxhole.

"Then it just stops."

And there was no further gunfire for the rest of the night. "Three hours to daylight," Dan recalled, "we sat there under flare light wondering if that was just a probe or if we were still under attack."

At daybreak, Charlie Company got their answer. At 3:00 AM, the Australian troops had opened fire on a VC night patrol. The Aussie with the Starlight scope had turned off the device. "The Starlight scope had to be used sparingly," said Dan, "because it's a battery hog and batteries were hard to get." Thus, the Australian rifleman was scanning his sector by noise, ambient light, and instinct. "He took a look down the road bed… *VC!* 100 yards on the far side of the road. These Aussies are on par with our Special Forces and they are bilingual. They speak Vietnamese, and he challenges the VC"—meaning that he called out into the night, masquerading as a comrade—"and they answer his challenge." He told them to come forward, and the unsuspecting VC took the bait. That's when the M-60 gunner opened fire.

But who had been firing the .50 cal?

As it turned out, the culprit had been an M113 Armored Personnel Carrier (APC) attached to the battalion's Bravo Company. "Company B's perimeter was right down the road from ours," said Dan, "and *nobody* had told the Aussies about B Company." Thus, when the Australian gunner opened fire with his M-60, the bullets went through the Viet Cong and impacted on Bravo Company's perimeter. The startled APC crew, thinking they were under attack, hurriedly returned fire—hence the unwelcomed .50 cal bullets strafing Crowley's foxhole.

"I walk over to the road to see what had happened [to the enemy scouts]," Dan continued. "There are 8 to 10 very dead VC lying in the road bed. One is dressed in a blue shirt and hat, the others in black PJs. The one in blue had an AK-47, the rest a mismash of SKSs, WWII carbines, and a couple of stick grenades. The M-60 did a job on the VC—they were well-perforated…and in a line like fallen dominoes." As Crowley inspected the last body in the line of fallen VC, the 1RAR sergeant came over to assess the damage inflicted by his mates.

"As we are standing there, he looks past and something catches his eye—it's a blood trail. I see it, too. The Australian sergeant with his M-16, and Crowley with his M-79, followed the trail some forty feet until it ended at a pile of brush and bamboo. Instantly, both men knew that they were tracking a wounded VC. "I'm covering the sergeant as he bends over and reaches into the brush pile, and pulls out a VC by the ankle." The VC soldier was unconscious, sporting three prominent wounds from the M-60. The Australian sergeant was about to shoot him until Lieutenant Coppolo waved him off, insisting that the VC be taken prisoner, treated for his wounds, and interrogated—for if the enemy scout regained consciousness, he might be a valuable source of intelligence. Coppolo and his driver, Specialist Le Beau, hoisted the ailing VC onto the back of their Jeep, intending to take him to the nearest medical LZ. However, the VC succumbed to his wounds en route to the MEDEVAC.

"Later that afternoon," said Dan, "the supply truck dropped off rations, equipment, and beer!," the latter of which they happily shared with the Australians. "Some had odd looking rifles [L1A1 Self-Loading Rifles to supplement their M-16s], others had cut-down M-60s—no bipods and shorter stocks. The Aussies were very professional in the way they carried themselves. I'm thankful they were on our side." Chuck Humphrey agreed: "It was obvious that they were well-trained," he said. Aside from interdicting the nighttime Viet Cong patrols, Humphrey remarked how enthusiastic the Australians had been during their tunnel rat missions. "They set up a mechanism in which they did a full mapping of the tunnel complex in and around Cu Chi."

Right before ROLLING STONE ended, Charlie Company welcomed the return of PFC Henrick, the unlucky soldier who'd been shot in the buttocks by an errant helicopter. He had made a full recovery from his wounds and, in celebration of his return, Henrick and his friend, PFC Lester, followed Joe Vargas's example by wrangling a pet monkey. "They have him on a chain," Dan remembered, "when they are driving the 5-ton dump truck...the monkey rides on the hood and he looks like a large Mack truck emblem." It was a comical sight, and this living hood ornament drew laughs from many a fellow trooper. These laughs abruptly stopped, however, when Henrick drove by the Battalion XO's Jeep. "The Major spotted the monkey on the hood," said Dan, "and when he got through with them [Henrick and Lester], Henrick had a little less ass to match his wound!"—meaning that Henrick had gotten a proverbial "ass-chewing" of epic proportions.

But Henrick's rotten luck wasn't done yet.

"Not long after this, we were convoying down to Di An, and one of the new guys had scored a .45 caliber grease gun—WWII type." The term "grease gun" referred to the M3 submachine gun used by Allied infantry units during World War II. Given its age, however, Dan noticed that the newbie's grease gun was problematic. It would jam and it had faulty magazines, thus rendering it a "single shot" weapon. "He jumped off the truck [at Di An], and it accidentally fired one round," said Dan, "hitting Henrick in the chest." Luckily, Henrick was wearing his ballistic vest, and the impact of the bullet left nothing more than a bruise. Still, the .45 ACP had hit Henrick with enough force to knock him to the ground. Less than a year in Vietnam, "and Henrick had already been shot twice by our own side...with no Purple Hearts for it all."

VC probes and friendly fire notwithstanding, Operation ROLLING STONE had been another tactical success for Charlie Company and the 1st Engineer Battalion. The new road from Ben

Cat to Highway 16 measured thirty feet wide; and the engineers finished the project in only *three* weeks. The rate of construction was so impressive that General Westmoreland wrote them a letter of commendation, stating:

> "The successful completion of this project, accomplished under hazardous conditions, exemplifies the importance of the engineer effort in Vietnam. Its ultimate effect on our future military operations, and on the Revolutionary Development program within the province, will be significant. This outstanding performance by the 1st Engineer Battalion is highly commendable."

By the end of ROLLING STONE, the 1st Engineer Battalion suffered three KIA and 29 wounded. Enemy losses totaled 148 killed.

On March 3, 1966, Charlie Company convoyed into Di An, while Delta Company rotated into Lai Khe to take their place. "Boy, was I glad to get out of Lai Khe," said Dan Crowley, "but then again, I had spent most of the last two months in the field with the Infantry on Operation CRIMP and ROLLING STONE doing sweeps, route clearing, and road building."

With Delta Company now settling into Lai Khe, Charlie Company took the former's mission of guarding and maintaining the infrastructure at Division Headquarters. As Dan Crowley described it, the company was now "in the rear with the gear." And, by most accounts, life was much easier than it had been at Lai Khe. "The VC were here, but not as active," he continued. "We still had snipers and mine layers." Di An had much less jungle vegetation and was dominated by plantation groves. According to Crowley, this meant no monkeys and less frequent mortar attacks. "Di An also had a few added benefits," he said—"our outhouse had screen doors, a mess hall that served C-Rations, showers…and tents with only a couple of holes in the roof,"—compared to the dozens of holes that peppered the Lai Khe tents. "The floors are PSP—Pierce Steel Planking, 5 inches, off the ground. Dropping anything valuable through the perforated flooring meant it was history. The rats loved it, and the snakes came looking for the rats. It was a dangerous neighborhood."

Charlie Company would spend the next few months at Di An. "It was during those months," said Chuck Humphrey, "that my platoon was in charge of vertical construction at the base camp. We were putting up buildings for office areas and barracks, all according to this great big book—the Engineer Field Manual—that gave you all the specifications of how to put up these buildings." However, for base construction, the Army didn't expect its engineers to accomplish the task alone.

"They hired Vietnamese laborers," said Chuck. "And a lot of them."

In fact, Chuck's platoon was responsible for a group of 50 laborers. "My platoon was supervising their work, day to day, for almost three months. And it's memorable because the bridge company of an engineer battalion has really big trucks for hauling the bridge equipment. So, they'd take those empty bridge trucks, they'd go down to the village, and load up all these Vietnamese civilians and bring them up to the base camp. They'd work all day, and at the end of the day, we had to pay them. So, I'd go get some money from someplace in the Division Headquarters, and I'd stand there and pay each of these workers. Then, they'd get on the truck and we'd take them home."

In between these construction projects, however, 3d Platoon found itself supporting Operation LAM SON II. It was a "pacification" mission intended to quell the VC along the western edge of the 1st Infantry Division's camp network. By May 1966, the Division Commander, General William DePuy, realized that local ARVN forces couldn't provide adequate security for the Division's rear areas. DePuy thus brokered a deal with the South Vietnamese government to provide a Revolutionary Development cadre (supported by elements of the 1st Infantry Division) to pacify the areas around Phu Loi and Di An. "The 1st Division would furnish an infantry battalion and a task force to provide psywar, intelligence, and other support," said Lieutenant Colonel Paul Gorman, whose own 1-26 Infantry battalion provided the maneuver "muscle" to the pacification mission.

"From June to September 1966," said Chuck, "my platoon was north of the Division Headquarters at a place called Phu Loi." As part of the ongoing pacification effort, Chuck and his men became provisional infantry, leading patrols beyond the perimeter at night. "It was my whole platoon," he said, "in other words, thirty soldiers. We would leave the perimeter right after dusk, and we'd get upwards of a quarter-mile away from the perimeter defense… and sit there all night hoping that nobody would discover us." Essentially, 3d Platoon was going outside the wire to act as human bait, hoping to draw fire from any VC lingering in the dark.

Not a particularly sound strategy - he thought.

But he had to follow orders.

"I can admit now that I felt as uneasy during those night patrols as any time I was over there. But fortunately, we didn't get attacked." Clearly, the job wasn't glamourous nor was it a high priority for the Division. Indeed, General DePuy was more focused on defeating the enemy's main force along Highway 13.

Meanwhile, back at Di An, the rest of Charlie Company carried on with a growing list of infrastructure projects. These included building a Division Chapel; mixing and pouring concrete; setting perimeter bunkers; laying minefields; and expanding the base's airfield.

Of these summer projects, however, emplacing minefields was the most memorable.

As Dan Crowley recalled: "The airstrip and helipad at Di An were being enlarged and Charlie Company was tasked with getting it done. We also had to lay a new minefield…it was all anti-personnel mines, a few thousand of them nicknamed 'Toe Poppers.' *It was up to me to arm them.*"

This minefield was intended to surround the Di An airstrip, protecting it from any prowling VC saboteurs. "The field was about twenty feet wide with parallel fences on both sides consisting of barbed wire"—with prominent "MINE" banners placed every few yards. "There is also a concertina fence running along the outside of it all." The Corps of Engineers had a tedious process for laying minefields. "It's recorded in a log book," said Dan, "along with a drawing and layout of where each mine is placed." To facilitate this process, the minelayer had to establish a reference point on the ground using a compass and transit. "Then you drive a stake in the ground," Dan continued. "Next, we use chains about 25 feet long," connected in series with shorter chains extending outward like an octopus connected to the mines..."and on each end, there was a brass ring." Each ring had a lettered brass tab—A, B, C, etc. "A drift pin is used to connect the ends of the chains together." Once the chains were connected, the mines were dug at pre-determined points along the links.

"Some team members would lay out the chains, others would dig holes for the mines. I was to follow, and arm the mines by removing the safety pins and covering them up." To accomplish this, Crowley had to crawl on his hands and knees, meticulously arming and concealing every landmine as he went along.

Dan's work was going smoothly...until one foolish comrade ruined it.

"As we were working," he said, "we noticed a GI approaching us on the outside of the perimeter." From his direction of travel, Dan could tell that this fellow GI was headed towards a bunker that had since been removed. "He didn't get the word." But then, Dan realized something peculiar about this wayward GI: to get to this part of the perimeter, he had to have crawled *through* the concertina wire.

Dan was perplexed.

How did he shimmy through the concertina?

Moreover, how had he emerged unscathed?

But this daring contortionist was about to have bigger problems—he was wandering straight into a minefield. "He stops at the barb wire fence," said Dan, "I'm like three feet away arming mines." The daredevil trooper certainly had the presence of mind to see what was happening in front of him. Suddenly, Major Rose, the battalion staff officer who was supervising the minelaying, yelled out:

"Go back and around - this is a minefield!"

The soldier, carrying a transistor radio, poncho liner, and his M-14, seemed to understand the Major's admonition. Curiously, the soldier then looked over at Crowley, as if to seek confirmation.

But Crowley simply told him: "Go back the way you came."

Dan went back to work, on his hands and knees, thinking nothing more of the situation. But he could still see the other soldier in his peripheral vision. "He pulls out a smoke and lights up." Moments later, however, Dan noticed that the barbed wire was *moving*.

"I turn around in time to see him already over the fence."

Dan was horrified.

He and Major Rose began screaming for the soldier to stop.

But it was too late…the foolish daredevil stepped on a landmine.

"And it worked," Crowley added.

Dan was only a few feet away when the mine detonated. "The blast blew me into a somersault." Dazed, dizzy, and barely conscious, Dan Crowley's first cohesive memory after the blast was the ringing in his ears.

"I'm covered in dirt, blood, bone, and boot,"—none of which was his.

Luckily, Dan hadn't been skewered by any shrapnel. The mine-running GI, however, wasn't so lucky. "He is missing one-half of his foot…and he is laying there. People are yelling to him not to move." Major Rose then ran over to the barely-lucid Dan Crowley. Despite having been catapulted by the blast, Crowley had somehow maintained his grip on the live mine he'd been holding. Rose took the mine from Crowley's hand and disarmed it.

"Now we have to remove him [the injured soldier] from the live mines," Dan continued. But this GI's hutzpah (and foolishness) continued unabated. "He sees his [cigarette] and he crawls to it over the armed mines."

Dan couldn't believe it.

This man had just lost his own foot…and he was more concerned with saving his *cigarette*? "He is so lucky that no more mines go off. We were able to tell where the other mines were located because the dirt was dry and fresh." While the psychotic mine runner was evacuated out of theater, Dan Crowley got checked by the medics and given the rest of the day off.

"He lost a foot;" said Dan, "I lost my hearing."

Although his hearing soon returned, Dan still wrestles with the memory of that horrific day. "Relentlessly clouding the quiet corners of my mind, triggering brutal nightmares, I still ask myself: Why did he cross the fence?"

Why indeed?

Was it simply bravado or arrogance? An undiagnosed mental disorder? Was he deliberately trying to injure himself? Was it a veiled suicide attempt?

Dan never knew.

Although the incident left him emotionally scarred, he had little choice but to carry on with his duties. "Next day, I was back to laying mines like nothing had happened," he said. "After the minefield was completed, I worked on building new bunkers and placing gas bombs [propellant napalm devices] in the perimeter."

Building these new bunkers had become a public health necessity—the pre-existing bunkers were "old and nasty," as Dan called them, and they'd become infested with vermin. For deployed soldiers, coexisting with rodents and insects was a fact of life. But these vermin were attacking the troops at night. For instance, one soldier in Charlie Company went to sleep after finishing his guard shift, only to be awakened by a gargantuan rat chewing through his cheek. "He had to get rabies shots... administered into his stomach," said Dan. "After that, I took to sleeping on the roof."

But, at times, even sleeping on the roof had its perils. One evening, while relaxing atop his bunker, Dan's relaxation was interrupted by the sudden burst of .50 caliber gunfire...but these rounds were landing *inside* the perimeter.

"What the hell is going on?" he wondered.

He hoped it wasn't another "friendly fire" incident, as had happened with the Australians months earlier.

"Those on top of the bunker dove into it," he said.

Fifty yards away, some tentmates who had been playing cards jumped from their tent, and ducked into a nearby slit trench. "These trenches had nasty green algae water mixed with mud,"

Dan recalled, "but when you are looking for cover, you can't be picky."

Crowley and others returned fire until they realized the anonymous .50 caliber gunfire had stopped. "Nobody was hit by the incoming," he said – just a few bruised ribs and sprained ankles from those who had piled into the trench. But the culprit behind the gunfire would shock him. "It's hard to imagine humping a .50 caliber on an ambush patrol, but that's exactly what had happened." As it turned out, there was an ROK (South Korean) Army camp adjacent to the US base camp—"and the Koreans had set up an ambush in front of our position without letting us know." Two Viet Cong had wandered into the ROK's path, and the ambush patrol opened fire.

"They started shooting at the VC," said Dan, "and the VC shot back."

During that melee, the Koreans' .50 caliber tracer rounds started landing in the US perimeter, "so we returned fire." The end result was two VC killed; one Korean KIA; and a US supply tent gone up in flames.

Emplacing gasoline bombs (propellant napalm) along the camp's perimeter was an equally-perilous task. "We would take 30-gallon or 55-gallon oil drums, cut the tops out of one, cut the ends off the other, and weld the two together," said Crowley. "Then, we'd mix the 'napa' powder with gas and it would gel up, making napalm. We'd place that drum at a 30-degree angle and cover it with sandbags, leaving one end open. At the other end, we would place a 'Willie Pete' grenade – M34 white phosphorous smoke grenade." White phosphorous was a highly-corrosive, highly-flammable compound that was designed to peel the flesh from an enemy soldier's bones. "Then," he continued, "we'd add a couple of blocks of TNT with an electric blasting cap and wire it back to the command bunker. Thank goodness we never had to use it. But if we did, it would make one hell of a crispy critter out of the VC."

Finally, in July 1966, Larry Blair took command of Charlie Company. Serendipitously, the 1st Engineer Battalion's Command

Sergeant Major had served with Larry in West Germany years earlier. Hoping that the Sergeant Major might persuade Lieutenant Colonel Kiernan, Larry approached the old NCO and said: "Sergeant Major, you gotta talk to him. You gotta talk to the old man here and get me a company. This will be the culmination of my training if I can lead a company in combat, you know?"

After all, Larry had been sent to Vietnam for that very purpose.

And thus far, he had commanded nothing except dead space at HQ.

"So, thanks to the Sergeant Major," Larry continued, "I got the job." A few days later, Larry Blair received the order to begin his change-of-command inventories, and take the guidon from Captain Zilenski.

Taking command, Larry was impressed by the caliber, resilience, and adaptability of his troops. When he took the helm at Charlie Company, the men were still committed to the construction project at Di An. "Our mission," said Larry, "was to get everybody in the Division out of tents before we could get out of ours." Thus, Charlie Company ensured that every unit at Di An was living inside a bunker before the engineers built their own.

"Most of the horizontal and vertical construction was on a pioneer basis," - i.e., crude and field-expedient—"but enough to get the job done." And although time-consuming, the process for erecting these buildings was surprisingly straightforward.

"We called them 'Sea Huts,' short for 'Southeast Asia Huts,'" Larry continued.

After laying a slab of concrete, the engineers would raise a 16x32-foot wooden frame on stilts nearly four feet off the ground. Every frame had a series of pre-cut, horizontal sidelight windows lining the upper and lower sections of the walls. These horizontal sidelights would then be fitted for wire mesh, allowing maximum air circulation while keeping out mosquitos and other unwanted insects. "And then it had a tin roof," Larry added, "just a corrugated tin roof." It might not stop an enemy mortar, but everyone agreed that the tin roof was better than a GI tent.

In between these construction projects, Larry marveled at

how seamlessly his engineers could fight as infantry on command. True, the engineers were *expected* to fight as provisional infantry as part of their doctrinal mission, but some units had more inertia than others.

Not so with Charlie Company.

"During the year that I was company commander," said Larry, "we were converted to infantry *three* times." These impromptu missions could vary in their size and scope, but they were typically *defensive* in nature. As Larry explained: "If an infantry battalion was given a mission to provide security for an artillery base… and that infantry battalion then got committed to a firefight, they [Division HQ] would take one or two engineer companies and say: 'You are now infantry; go provide security for that artillery unit.' So, we weren't like an assault infantry. Typically, we were in more of a defensive mode…setting up blocking positions or assuming perimeter duties…because we weren't quite as mobile as the infantry, but we did have a lot of heavier weapons."

As the Di An mission continued, Crowley remembered that "mail call" was one of the few things that kept his sanity. "Most times, mail call was held at the evening formation by the Company Clerk. I was not expecting any, but my name was the last one called that evening."

Receiving his letter, however, Dan noticed something odd.

"The envelope looked official."

This was not the informal stationery that he typically received. Then, he saw the return address:

"Rear Admiral BB Crowley, Ret. (TLF-RB6) Mayport, FL."

It was an inside joke from Dan's older brother. "The BB stands for 'Big Bologna' and TLF -RB6 stood for 'Twin Lakes Fleet, 5 row boats.'" Still, the "Admiral Crowley" return address had been enough to get the First Sergeant's attention. When the First Sergeant called him over, asking about the letter, Dan admitted: "I tried not to bust a gut laughing in answering him. I told him the letter was from my uncle. He thinks I have an Admiral in my family but it's just my big brother being a jokester." In reality, the elder Crowley was a prior-enlisted sailor aboard a Navy destroyer.

"I guess you could say the joke was on the First Sergeant and the whole orderly room. This gave me a real good laugh to sleep on that night."

The following morning, Dan Crowley started the day with his semi-regular routine of dawn minesweeping. "Most mornings at Di An," he recalled, "I would be on a mine sweep mission to clear the road outside of the base camp." The Viet Cong operated mostly at night, using the cover of darkness to lay mines and IEDs. American combat engineers, therefore, had to sweep the roads for any ordnance placed overnight. Minesweeping teams were typically led by an NCO, accompanied by "three men working the metal detectors…three more men following with bayonets to probe detector hits, and a demo man to blow in place anything that looked suspicious."

After clearing roads in the morning, Dan's demo team would be assigned any number of random tasks that the battalion generated on a given day. Of these tasks and work details, Dan's favorite was driving the 5-ton dump truck. "Most of our runs were to the gravel and laterite pits at Long Binh," he said, "over an hour away." For every convoy, however, the vehicle had to have at least *two* soldiers in the cab—a driver and a so-called "vehicle commander." Often, the vehicle commander had no real authority over the driver. Unless the vehicle was assigned to a key leader with a designated driver, this two-man setup was essentially a "buddy system" implemented for safety's sake. But, as with many regulations, the enlisted men pushed the boundaries…and learned that this was a rule they could safely afford to break.

As Dan recalled: "Riding shotgun from Point A to Point B wasn't necessary. There was too much to see and do. There were bars on both sides of the road for miles; you could work up a thirst with young ladies selling cold beer. One of our hang outs was the *Whiskey-a-Go-Go*,"—a roadside bar known for its fine lagers and fast women. The enlisted men in Charlie Company soon devised a system wherein, on their way to the laterite pits, the "shotgun rider" (vehicle commander) would be dropped off at the *Whiskey-a-Go-Go*, and get picked up on the return trip.

During one trip, Dan Crowley arrived at the bar…only to

notice another Charlie Company truck (designated "C-17") parked behind the building.

"This was not good," he said.

These trucks were supposed to *drop off* and *pick up* the shotgun riders during their round trips to the laterite pit. But if a truck (including both of its crew) stopped at the bar, it would delay their anticipated arrival times, and thus draw suspicion from Charlie Company's leadership.

"Our scheme had been working fine with no problems," he recalled, but now it seemed that the jig was up. The "C-17" designation meant that the truck belonged to 1st Platoon (the 7th-listed vehicle in the platoon's inventory). "As I entered the bar, I see C-17's driver. He has a girl sitting on his lap while he's drinking a beer. His shotgun [i.e. vehicle commander] is nowhere to be seen, probably drowning in a sea of love."

Little did they know, however, Captain Blair had been keeping track of the daily load deliveries. And his quick audit revealed that C-17 had not yet delivered. Realizing that the truck crew had likely broken down or taken a romantic detour, Larry Blair went hunting for C-17.

"I'm drinking a cold one," Crowley recalled, "when someone yells 'Fire in the Hole!,'"—meaning that they had just seen Captain Blair.

"Oh Shit!" cried Dan.

"Suddenly the backdoor became very crowded," he continued—and Dan Crowley was among the handful of young GIs running from the *Whiskey-a-Go-Go*. Luckily, Dan's truck and driver were inbound at that exact moment, and the young Crowley jumped into the cab as the irate Larry Blair went prowling through the bar, looking for his missing GIs.

Luckily, Dan escaped undetected, but the C-17 crew wasn't so fortunate. But when Blair confronted the wayward crew, they claimed that the trucks' air filter had clogged, and they needed to stay with the truck until help came. It was just "coincidental" that the breakdown happened near *Whiskey-a-Go-Go*.

Soon, Dan Crowley and his demo team were back in the fight, supporting infantry patrols. Charlie Company was still at Di An, but the increasing number of search-and-destroy missions had likewise increased the demand for combat engineer support. "I had been in the field with the infantry for a few days and when I returned to base, I saw some of our dump trucks had been armored with steel plating bolted to the doors." This steel plating stood at 1.5 inches thick, measuring 4x6 feet in dimension. "This was a bad idea," Crowley admitted. It impacted the crew's field of vision, left no room for side mirrors, "and the hinges were insufficient to hold the doors, so they soon fell off." Moreover, Dan wondered how anyone could back up a truck without the aid of side mirrors?

"I was lucky I didn't have to drive one."

Indeed, by this time he was no longer making runs to the laterite pits. Now, he had graduated to building the Division Chapel. Together with his friend PFC Dave Smith, and Staff Sergeant Ramirez, Dan Crowley became part of the supervisory/assistance crew overseeing fifteen Vietnamese laborers and carpenters. With a seating capacity of 220, the Division Chapel was the "crowning achievement" of the 1st Engineers' construction projects at Di An. During this time, Crowley also received a promotion from PFC to Specialist.

Although Charlie Company continued supporting local combat operations, "increasing emphasis was placed on base camp development at Di An." For example, Charlie Company alleviated the camp's perennial drainage problems by laying 800 feet of culvert. Aside from their developmental work on the airstrip and the nearly four-dozen camp buildings, Charlie Company also poured an astounding 85,000 cubic feet of concrete.

"Some nights we would mix concrete by hand and pour billet pads for the base camp," said Dan. "Other nights we played cards and, if we were lucky, drink beer…which was always in short supply." Luckily for Dan Crowley, however, his tentmates included two "old-timer" sergeants—Snyder and Gabbard—who were, respectively, the senior grader and bulldozer operators in

Charlie Company. Both men were "lifers" who had been in the Army for several years, but never advanced beyond the rank of buck sergeant. Still, the pair had been in the Army for so long that they now had "connections" and "friends in high places" far beyond Charlie Company. As Dan recalled: "They got word through the 'Lifer Telegraph' that an old pal of theirs was working on the docks in Saigon and he had access to beer!"

Snyder and Gabbard somehow talked the First Sergeant into letting them use an M35 two-and-a-half-ton truck (the venerated "Deuce-and-a-half" cargo truck) to check out the beer lead. "They hit the mother lode—pallets of beer," Dan recalled, "and a small hill of broken cases." Their dockworker friend simply told them to "take what they want" and left the pair unattended.

Snyder and Gabbard then commandeered a nearby forklift.

"They put two pallets on the truck," and in between the open space of the pallets, they stuffed as many single cans and broken cases as they could fit. "Needless to say," Dan chuckled, "they were heroes when they returned."

A few days later, Crowley was told he would be going on an ambush patrol—"so, I loaded extra ammo and grenades." Ambush patrols, by their very nature, were a game of luck. On a pre-selected piece of terrain (usually along the side of a well-travelled road), a group of soldiers would lie in wait, hidden from view, and with their weapons at the ready. If an enemy patrol or convoy came into view, the ambush would open fire on that unsuspecting enemy. Of course, the success of an ambush patrol depended on whether or not the enemy came into its path. Sometimes, an ambush patrol might lie in wait for hours…and make no enemy contact. On other occasions, an ambush might intercept multiple targets on the same day.

Today's patrol, however, would have an unexpected twist.

"We headed out the gate going towards Long Binh," Dan recalled. "There were twelve of us - a 2nd Lieutenant and eleven enlisted men traveling aboard two dump trucks, with a Jeep in between." The convoy drove nearly half a mile before turning onto a one-lane dirt road. "We passed a small ARVN outpost," he continued, "and the road began to turn to shit."

Now he was worried.

They were travelling down an unfamiliar, unimproved road that hadn't been featured in any of the recent minesweeping missions.

"Didn't our 1st Platoon do this back in January?"—he asked himself rhetorically.

"Heading down a road that hadn't been swept?"

He was referring to the ill-fated route reconnaissance wherein Jay Franz had been wounded, and Sergeant Carroll killed in action. "At least I'm standing on a double layer of sandbags in the bed of a dump truck in case we hit a mine," he thought to himself.

Sadly, his premonition was about to come true.

Crowley was riding in the lead truck, with the Jeep following close behind. "I'm looking back at the Jeep when I see a flash, a boom, and clouds of dust and dirt." An IED exploded mere inches in front of the Jeep. A moment later, a smattering of dust and dirt came raining down onto the troops in the backbed of the leading dump truck. The Jeep's driver, Specialist Joe Brown, reflexively cranked the steering wheel hard right…which nearly rolled the vehicle over. "Dirt covered the hood," Dan recalled, "and his eyes were as big as saucers." Shaken, but otherwise unharmed, the Jeep crew dismounted to inspect the road damage.

"The mine was homemade with a command detonator," and Crowley likened it to a tin can with a kite's tail—describing the command wire that trailed off into the distance. "It reminded me of when I was a kid," he continued. "I would take two tin cans, one a little larger than the other…cut tops out of both cans, and I'd punch a hole in the lid of the smaller can. Then, I'd push a firecracker fuse up through the hole." After adding a few ounces of water to the larger can, Crowley would insert the smaller can, then light the fuse. "The small can would launch [like a rocket]."

Snapping himself back to the present moment, Crowley scanned the horizon looking for the perpetrator behind the IED. He couldn't identify the bomber, but Dan knew that this faceless VC had likely reported the Americans' presence to the local cadre. The trucks and Jeep sped back to Di An, leaving Crowley and the others to set up their ambush. "It was uneventful," Crowley admitted, "so, the next morning, we walked back to base camp."

In all likelihood, the VC bomber had warned his brethren to stay away from the area, sensing that the Americans were setting up an ambush patrol.

A few days later, Crowley and a fellow GI, Specialist Morris, were inbound to Di An from Lai Khe aboard a Huey helicopter. "As we near Di An," he recalled, "we see a cloud of dust and a C-47 Chinook is about to land on the parade ground." With its iconic tandem rotor blades, the Chinook was a heavy-lift helicopter that had been adapted into a variety of aerial support roles. Although he was impressed by the sight of a Chinook, Dan thought nothing more of the exotic helicopter until his First Sergeant told him to report to the parade ground. As it turned out, 1st Platoon would execute their next few aerial insertions from a Chinook. To accommodate the Division's expanding footprint across the jungle, "we had to practice using a Chinook," said Dan, "to cut landing zones in out-of-the-way places."

In fact, the engineers had to use *two* Chinooks for LZ clearing—"one had mostly troops; the other had a sling load of demo, chain saws, gas, water, and hand tools." The Chinooks would fly the engineers to a hole in the jungle—"a bomb crater, usually," said Crowley—sending them down a rope ladder with instructions to make that hole into a serviceable LZ. Dan recalled that even when the Chinooks hovered at low altitude, their airframes would shake, rattle, and make so much noise, he'd wonder:

"Shit, is this bird going to hold together?!"

But two weeks later, these rickety Chinooks would deliver Dan's platoon into one of the vilest battles of 1966.

CHAPTER 5
FIELDS OF FIRE

By August 1966, the 1st Engineer Battalion was approaching the one-year anniversary of its deployment. Earlier that year, President Johnson announced that the standard tour of duty for all troops in Vietnam would be twelve months (except the Marines who were given thirteen-month rotations). Although this policy was intended to prevent battle fatigue, it robbed every unit of its most experienced troops. Indeed, those who had learned to fight and survive in Vietnam would be going home, taking with them all their experience and expertise. Replacing them would be a horde of new draftees, the training and quality of which deteriorated as the war dragged on and the Army scrambled for bodies.

Although the 1st Engineers had deployed as a singular unit, they would re-deploy as individuals. Trying to maintain a cohesive unit in the midst of high turnover rates was challenging. By the end of the war, one Army general remarked that: "we had the spectacle of officers standing up in the morning in front of squads, platoons, and soldiers whom they didn't know, and who didn't know them, or know one another. The officers didn't know the soldiers, the soldiers didn't know the officers, and they were supposed to go out and fight a battle that morning." But before the 1st Engineer Battalion began suffering under these HR policies, the unit would test its resolve at the Battle of Bong Trang.

Dubbed "Operation AMARILLO," the battle began as a simple route clearance mission along Highway 16 to facilitate the "much-needed resupply convoy from Di An to Phuoc Vinh." The 1st Infantry Division's 1st Brigade had established two artillery bases for the occasion, and placed an infantry battalion at each site. The 1st Battalion, 2d Infantry Regiment (1-2 Infantry) stood at Firebase 1, securing the road between Song Be and Tan Binh. Meanwhile, 1st Battalion, 26th Infantry (1-26 Infantry) occupied Firebase 2, along the southern edges of Tan Binh. Division Headquarters subsequently ordered the 1st Engineers to provide combat support for all 1st Brigade operations in the area.

From the condition of Highway 16, however, the engineers anticipated spending most of their time on road repairs. "The road was in poor condition and two culverts had been blown." Still, the mission was being touted as just another "ordinary road clearing job."

But Charlie Company would soon discover that this mission was "anything but ordinary."

On the morning of August 24, 1966, Dan Crowley recounted: "I was told to get my gear and report to the orderly room ASAP. I was informed that a three-man team would be going on a route clearing mission for Operation AMARILLO," along with the 1-4 Cavalry Squadron. Crowley enjoyed tagging up with 1-4 Cavalry; they were among the few mechanized units that hadn't been forced to leave their armored vehicles behind at Fort Riley. Thus, Crowley anticipated that his team would be riding atop an APC or tank at some point during the mission.

"Specialist Parish would be in charge of the team," said Dan, "Specialist Colon would have the mine detector," while Crowley himself would carry the demolition tools. His packing list was the same as it had been for every other mission: a Claymore mine; ten and one-half blocks of TNT; two blocks of C4 explosives; blasting caps; time fuse; det cord; more than a dozen rounds for his M-79 grenade launcher; two hand grenades, and a Colt .45.

"We were taken to the airstrip for a hop to Lai Khe," he continued. "They put us aboard a caribou…a bush plane made for short take offs and landings…what a ride!" Indeed, as the pilot throttled the plane's engine, the ensuing vibrations made it seem as though the plane would fall apart before it left the runway. "We rolled down the strip like a race car…then straight up," said Dan. "Coming into Lai Khe was even more exciting." The airstrip was short and Dan characterized their landing as more of a "controlled crash."

Coming off the airstrip, Dan and his team met up with C Troop, 1-4 Cavalry. "For some reason," he said, "they split up our team"—placing Specialist Colon into one APC, while sending Crowley and Parish into another APC. "It was really crowded [in the vehicle]," said Dan, "but we got to ride up top." The APC caravan departed Lai Khe heading south towards Ben Cat along Highway 13. "We peeled off Route 13 and went cross country," Dan continued. "Having spent three months of my tour at Lai Khe, I knew we would cross the road that the 1st Engineers built in February of 1966"—the infamous Route Orange, otherwise known as the "Rolling Stone Road," where he and his comrades first met the Australian commandos.

"As fast as we built that road, the VC would mine it."

When the convoy shuddered to a halt along Route Orange, Crowley asked the officer-in-charge (a young lieutenant) if they intended to clear the route.

"No," replied the lieutenant.

But Dan had noticed a problem.

"With two tanks and nine APCs set to go down the road, I let the lieutenant know about the previous problems we had with VC mines." Moreover, Dan had noticed that the South Vietnamese civilians had been staying away from Route Orange—a telltale sign that the VC had been active along the road. Still, the lieutenant did nothing to halt those tanks and APCs…until the leading tank drove over a landmine.

"BOOM! It blew off the track and a road wheel."

"Now we had to wait for a track retriever to come and tow it back to Lai Khe."

Later that afternoon, the convoy set up a perimeter for the night near Highway 16. Shortly after sunset, one of the line companies from 1-2 Infantry dispatched a Long-Range Reconnaissance Patrol (LRP) into the night, scouting the area for any VC. But after several hours with no enemy contact, the LRP hunkered down for the night, setting up their patrol base north of Bong Trang.

The next morning, however, the LRP awoke to find that they had encamped themselves *right in the middle of a Viet Cong base*. Within minutes of daybreak, the VC discovered the wayward American patrol, and began attacking from all sides.

As Dan recalled: "That's when all hell broke loose."

Panicking, the 15-man LRP dug in and returned fire while the patrol leader made a desperate plea for help over the radio. In response, their company commander, Captain Bill Mullen, mobilized the rest of his troops to rescue the beleaguered patrol. "M-16s; M-60s; AK-47s; grenades—everything was going off."

Although Dan's vehicle was quite a distance from the perilous patrol, he could hear the unmistakable sound of Allied and Communist ordnance trading fire with each other. Even more immediate were the frantic voices over the radio, all of which were piping into the M113's internal speaker.

Captain Mullen quickly mounted his remaining men atop the vehicles in 1-4 Cavalry, and the convoy sped off into the fray. "I was amazed how fast the cavalry and infantry responded," said Crowley. "We rode to the 'sound of the guns' with a tank up front, nine APCs in column, and a tank in the rear with infantry on board."

As Captain Mullen's troops (along with 1-4 Cavalry) hurtled down the road, the 1st Brigade commander, Colonel Sid Berry, ordered the nearby 1-26 Infantry battalion to send a company to reinforce Mullen's rescue effort. Division had now declared this a "tactical emergency"—meaning that all available ground units and helicopters were being diverted to help 1st Brigade recover the lost patrol.

From his position atop the APC, Dan Crowley could see that the leading tank was "busting a trail for us to follow." To facilitate rescuing their comrades, the convoy was now going cross-country, blazing the quickest trail possible into the LRP's location. "We went over a trench and a berm," said Dan, before the infantry dismounted and ran into the sound of enemy fire. "Parish and I stayed on the APC, with him on one side, I on the other side, and the track commander behind the .50 caliber."

Soon, the VC turned their attention away from the lost patrol, and prepared to engage the incoming rescue force. "With all kinds of explosions going off," Dan recalled, "I could see in between the open spaces, under tree canopies…bunkers, trenches and berms." By this time, Captain Mullen's troops were heavily engaged with the Viet Cong…but the APCs (including Crowley's) remained untouched.

"It went quiet for a moment," said Dan, "then the sh*t hit the fan."

As the sound of enemy bullets whipped past the APC, Crowley and his crew began returning fire. Their .50 caliber machine gun was the first to answer the call. "The track commander was smoking that thing." Crowley himself, meanwhile, began pumping grenades in rapid succession from the barrel of his M79.

Within moments, however, the VC got close enough to return the favor.

"Lots of hand grenades started flying," he said, "with two landing on top of the APC next to ours"—about 20-25 feet away. Luckily, the crew of that APC had shut the vehicle's hatches prior to the battle. Thus, the grenade had no effect on the vehicle or its crew.

The items hanging from the outside of the vehicle, however, weren't so lucky.

"Backpacks, water cans, cases of rations and ammo vanished… just gone." To make matters worse, the next incoming grenade was headed right towards Dan Crowley's APC. "It was a stick grenade…they had a bunch of those damn things.[9]" Despite their ubiquity, however, Crowley admitted that these "stick grenades" were comparatively weak. Although not necessarily as lethal (or aerodynamic) as the standard US grenade, they were still dangerous…and they could easily maim an Allied soldier at

---

[9] Allied troops often referred to these grenades as "potato mashers," having a strong resemblance to the popular kitchen utensil.

close distances. "That stick grenade hit the side of our .50 cal, and ended up next to our track commander's hatch."

Terrified, the APC commander ducked inside the vehicle.

But it was too late.

"The blast blew out the periscope window," Dan recalled, "giving him a face full of glass." Crowley and Parish, meanwhile, had ducked down into the rear hatch of the vehicle—"quickly avoiding KIA status," he added wryly.

Dazed and confused from the explosion, Parish and Crowley shimmied over to the disabled track commander. He was still alive, but his face was bloodied…peppered by shards of broken glass. "All three of us had concussions from that grenade," said Dan. Parish rendered First Aid to the wounded APC commander, while cleaning up the shards of glass scattered along the vehicle's floor. After regaining some of his lucidity, the APC commander ordered Crowley to man the .50 caliber machine gun.

Meanwhile, two additional US infantry battalions had been ordered to join the fight. "The 1st Battalion, 16th Infantry [1-16 Infantry] was airlifted from Lai Khe," and began attacking eastward. "By early afternoon, the VC were apparently breaking contact and moving west." The Americans responded with a volley of air and artillery strikes, hoping to disrupt the enemy's retreat. Soon thereafter, 2-28 Infantry was airlifted to set up blocking positions to the north. From the radio traffic, it was clear that the VC were operating as a battalion-sized element—and were thus outnumbered 3 to 1.

But as Dan Crowley soon discovered, the enemy's "withdrawal" would be a hellacious "fighting retreat"—they would expend every last round of ammunition as they fled from the field. In fact, as Crowley settled himself onto the .50 caliber machine gun, he could see the Viet Cong giving their "last hurrah" as they faded back into the woods. But although the Viet Cong were fading from this sector of the battlefield, he could see the damage they had done. "The tank took an RPG hit," he recalled, "and I could see two APCs on fire." His own APC, meanwhile, had overheated—the engine stalled from having idled too long.

As Dan continued scanning the area with his .50 caliber machine gun, the Viet Cong regrouped and counterattacked north along the forward edge of 1-2 Infantry's position. At the time, 1-2 Infantry's battalion commander was on leave in Hong Kong, and the executive officer, Major Clark, was nominally commanding the unit. "Major Clark went forward to take command at the scene of the battle and was killed almost instantly." Colonel Sid Berry, the 1st Brigade commander, then arrived on the scene to restore order, whereupon he encountered a critically-wounded Captain Mullen. The young company commander had been shot in the leg, and although his troops were holding strong, they were being badly shot up by one VC attack after another. "The whole area was swept by enemy fire, including some sniper fire from up in the trees."

As Mullen was briefing Berry on the situation, however, the young captain took a bullet to the chest. Mullen himself recalled that it was a superficial wound that had barely penetrated him, but the force was enough to knock Mullen over, and facilitate his evacuation from the battlefield. Berry ordered him into the nearest medevac vehicle, and sent a protesting Mullen on his way to the field hospital.

Shortly thereafter, at around 4:30 PM, a sudden blast of noise and movement caught Dan Crowley's attention. "I saw movement up through the canopy and the noise of an Air Force rescue helicopter"—a HH-43 Huskie. "It took a hit and crashed 100 feet in front of our APC." As it turned out, the helicopter had been struggling to stay aloft after sustaining a direct hit from a 57mm recoilless rifle from the center of the battle area. Over the next few hours, "troops strung out along the chopper to protect it." However, due to the thick jungle and heavy fighting, "the chopper could not be extracted until late the next afternoon."

As the battle shifted away from Crowley's sector, he recalled that 1-4 Cavalry withdrew later that afternoon. "There were only two or three APCs left running," he said. "The tank was going to be abandoned and blown up, but that didn't happen—instead they just stripped off its machine guns." Coming off the line,

Dan recalled: "We got to a cleared area and set up a perimeter for the night." As the 1-4 Cavalry troops consolidated within the perimeter, Crowley went looking for Specialist Colon, as he had been placed in a different APC prior to the battle. "He was nowhere to be found," said Dan, "and I was told his track had been hit and caught fire." As Crowley went looking for Colon, however, he spied an APC being unloaded with weapons collected from the American dead and wounded. "There were a lot of M-16s, shotguns, M-60s, and .50 calibers. Some of the M-16s had bullet holes and blood."

Walking past the APC, Crowley noticed two soldiers carrying a .50 caliber machine gun down the back ramp—"one had a hold of the barrel, the other had the handles." Dan continued walking, trying to avert his gaze from the mass of bloodied and pockmarked weapons. But he wasn't more than ten paces beyond the APC when he heard the .50 caliber go off.

Dan spun around with a startle.

The soldier who had been holding the barrel was now lying on his back.

A single bullet had been left in the feed tray, and the soldier carrying the opposite end of the machine gun had accidentally discharged the weapon. For the poor soldier carrying the barrel, the bullet's force of impact had been enough to launch him *15 feet* away from where he had been standing. From among the frantic cries of "Medic!" Dan noticed that the bullet had gone "clean through" the young man's shoulder.

Exhausted, deflated, and running on adrenaline, Dan Crowley staggered back to the spot where Parish was digging their foxhole. "I sat down next to the foxhole and started to shake, cry, and puke—I was scared."

And who could blame him?

"I grabbed the entrenching tool and started to dig with complete abandonment." Parish eventually told him to stop— "the way I was going, we wouldn't be able to see out of it if we had to use it."

That night, Dan recalled that: "It never got dark…with one flare after another…impacting artillery and repetitive mortar rounds exploding on the VC base camp. In the morning, the VC were gone."

But one question still lingered on his mind.
Where was Colon?
Had he been killed? Wounded? Captured?

Dan later found out that Colon had been medically evacuated, but no one would provide the details of his condition. "It took 35 years to find out what happened," said Dan—"grenade blast almost took his arm off." He spent the next ten months at Fort Sam Houston. "This mission…changed our lives forever."

Although the enemy faded away from Crowley's sector of the battlefield, the firefight raged on into that following morning, August 26, in 1-2 Infantry's area. Colonel Sid Berry, for his part, had taken control of the beleaguered battalions—1-2; 1-16; and 2-28. He had been trying to stem the tide of confusion and, despite the hail of gunfire, he had been moving amongst the units, locating leaders and restoring a coherent command structure. The tides turned savagely against the Viet Cong, however, when 1-26 Infantry arrived on the scene.

With that additional manpower, the American line units positioned themselves to envelop the diehard VC. Holding the enemy during the hours of darkness, the Americans renewed their assault on that morning of August 26. Sid Berry decided that a napalm strike would be the most effective means to drive the VC out of their fortifications. At dawn, 1-26 Infantry marked the enemy position with smoke grenades, broadcasting the VC's location for the incoming napalm strikes. The inbound fast-movers dropped more than twenty cans of napalm onto the now withering enemy. After Sid Berry decided to call off any further airstrikes, 1-26 Infantry assaulted through the now-cleared objective. When they recovered the lost LRP team from Captain Mullen's C Company, they discovered that only nine of the original 15-man patrol had survived.

Throughout the battle of August 25-26, Charlie Company (and the rest of the 1st Engineers) suffered only a handful of casualties, most of whom were in the Battalion Headquarters

Company. These casualties, though tragic, were minor compared to the other units who had fought to destroy the enemy battalion and rescue the lost LRP. The heavily-battered 1-2 Infantry, for example, lost eighteen soldiers with an additional 98 wounded.

As it turned out, the Americans had been fighting the Viet Cong's Phu Loi Battalion—a well-equipped cadre of hardened Communist fighters. Reflecting on the engagement, however, Charlie Company's performance highlighted the versatility they espoused as combat engineers.

They had fought as provisional infantry.

In the days that followed, the 1st Engineers were "called upon to destroy tunnel complexes discovered by the infantry." They continued to function as infantrymen until August 28, when the battalion resumed its regular engineering mission. "Jungle clearing, road and bridge repair, and cutting landing zones kept the engineers busy" until the official end of Operation AMARILLO on August 31, 1966.

Throughout the battle, the US tallied more than 100 confirmed enemy killed. Friendly casualties totaled 41 KIA, 34 of which were killed during the opening rounds of August 25. The US declared the operation a victory but, as Sid Berry added: "We thrashed our way almost blindly into the enemy's base camp and fought him on his home ground under conditions favorable to him." Still, the Americans had taken the field and decimated the Viet Cong.

The following month, however, Charlie Company steeled itself for another mission along the most dangerous highway in Vietnam.

Sergeant First Class Scroggins, the 3d Platoon Sergeant. As Chuck Humphrey's platoon sergeant, Scroggins provided a steady-handed tactical presence throughout Charlie Company's initial tour. Scroggins was wounded during an ambush along Highway 13, whereafter he spent six weeks in recovery.

Dan Crowley poses in front of the tactical dump truck, designated C-17.

Sergeant Paul Jaroun (left) and Sergeant Ray Griffin (right), both of 1st Platoon, 1966. Jaroun was one of Crowley's squad leaders. Griffin was among the most well-respected junior NCOs in the company.

Dan Crowley marvels in front of the crater left by a VC mine.

3d Platoon's tactical dump trucks. Although ostensibly for hauling laterite and various building materials, these trucks became general purpose reconnaissance vehicles for the duration of Charlie Company's deployment.

Repairing a bridge at Ben Cat, 1965. Among their regular duties, the combat engineers often built and/or repaired bridges, roads, and other points of infrastructure.

Minesweeping. When not repairing the local infrastructure, Charlie Company often conducted route reconnaissance missions. Whether mounted or dismounted, these route recon missions typically included minesweepers who would use commercial metal detectors. Upon locating these mines, men like Dan Crowley would detonate the mine in place.

Entrance to Charlie Company's encampment at Di An.

Celebrating Christmas 1965 and New Year's 1966. Soldiers of the 1st Infantry Division decorate the tank bulldozer blades for the holiday season.

M48-based tank bulldozer at Trung Lap. These armored dozers were among the few mechanized assets that the 1st Infantry Division was allowed to bring on their deployment to Vietnam.

Dan Crowley (driver side) and Dan Smith (passenger side) enjoy a photo op while driving a Jeep.

Specialist Parish (center), who accompanied Dan Crowley into the Battle of Bong Trang - August 25, 1966.

The so-called "Boom-Boom Girls" who frequented the local GI bars in and around Saigon.

Captain Larry Blair (left) confers with Sergeant Major Bird (center) and the Battalion Commander (right), LTC Joe Kiernan, prior to a mission, 1966. Kiernan was the top graduate in his West Point Class of 1948. Tragically, Kiernan was killed in June 1967 when his helicopter crashed due an avionics malfunction.

Sergeants Snyder and Gabbard. These older NCOs were a pair of Army "lifers," meaning they had been serving longer than most of the soldiers in Charlie Company…yet, they had never advanced beyond the rank of buck sergeant. Still, their long-standing service had yielded them numerous "connections" to "friends in high places" who could facilitate favors. Snyder and Gabbard cashed in one such favor by delivering a commandeered shipment of beer to their comrades in the field.

Specialist Watson (left) and Specialist Deal (right). Watson accompanied Dan Crowley on the route clearance mission wherein Crowley detonated an enemy mine moments before a command helicopter attempted to land in the area. Specialist Deal was another one of Crowley's demolition buddies. Deal had the unfortunate distinction of being bitten by a jungle rat in his sleep… necessitating a booster shot from the local field hospital.

The 1st Engineer Battalion completes work on a Bailey Bridge. These rapidly-deployable truss bridges were a common staple among engineer units, as they could be built quickly and could accommodate a wide variety of vehicular traffic.

Specialist James Januska deploys his mine detector on another route clearing mission along Highway 16 as Operation AMARILLO gets underway.

Men of Charlie Company descend from helicopter-mounted rope ladders into the jungle north of Suoi Da. These combat engineers are preparing to clear a landing zone for follow-on forces in support of Operation ATTLEBORO.

Preparing to turn the area into a landing zone, Dan Crowley uses a chainsaw to level tree stumps during Operation ATTLEBORO.

Dan Crowley on his post-deployment leave in Florida, January 1967.

Dan Crowley visits the Vietnam Veterans Memorial in Washington, DC. 2007. Commonly known as "The Wall," this memorial chronologically lists the names of all 58,318 Americans who gave their lives in Vietnam.

CHAPTER 6
THUNDER ROAD

As August gave way to September, Dan Crowley once again found himself on convoy duty—riding shotgun aboard a dump truck. One September morning, which he flippantly described as "another day in paradise," Dan was part of a four-truck convoy headed to the Long Binh gravel pit.

"When we arrive, I see a million-dollar rock crusher sitting idle."

That was odd—he thought to himself.

Why would such an expensive piece of equipment be standing unused?

As the truck rumbled closer to the gravel pit, however, Dan found his answer. The rock crusher's crewmen had been given a partial holiday...courtesy of the Army's criminal justice system. "Close by are about 20 GI prisoners with sledgehammers," said Dan, "guarded by half a dozen MPs with 12-gauge shotguns." These "GI prisoners" were rank-and-file soldiers who had been apprehended by the Military Police for various misdeeds—ranging from insubordination to public intoxication to attempted desertion. Now in custody, these GIs had been sentenced to hard labor as part of their penance. As such, the regular rock crusher crewmen had been told to "take a break" while the miscreant soldiers performed their tasks manually. "Talk about being sentenced to hard labor," Dan marveled—"turning large rock into gravel is about as hard as it can get!"

After loading up his dump truck with the requisite gravel, Dan and his driver headed to a bridge site on Route 16, just south of Tan Uyen. "The existing Vietnamese bridge, weakened by previous heavily laden convoys, was failing rapidly." Dubbed "Operation LONGVIEW," the 1st Engineers were tasked to replace the existing bridge by two 90-foot "Bailey Bridges" supported by one intermediate pile bent. The Bailey Bridge was a scalable, prefabricated truss bridge that had been in use among the Anglophone armies since World War II.

"When we arrived," Dan continued, "we were told to park the trucks and lend a hand carrying bridge panels." As Crowley remembered, panels were about "10 feet long and 6 feet high… very heavy and awkward, and it takes eight people to carry them." The panels had male and female ends, connected with four-inch drift pins inserted via sledgehammer. These scalable bridges were "built on the roadbed and pushed on rollers over the opening," said Dan. "It's a balancing act with more weight on the roadbed than over the opening, otherwise it would topple into the Song Be River."

All told, he likened the Bailey Bridge to a "giant erector set."

The bridge-building operation was unique because it was the battalion's first attempt at using a piledriver in Vietnam. "In no time at all, the bridge was completed," he said. "We got the job done. No Mission too difficult for the 1st Engineers."

But, for Dan Crowley, the signs of battle fatigue were beginning to show. By October 1966, he had been in Vietnam for ten months. He'd seen more death, destruction, and questionable decision-making than he had bargained for. And although his physical endurance remained strong, his patience was wearing thin—especially for arrogant, foolhardy leaders.

Such was the case when Crowley had to force-detonate a mine in front of his battalion command team. That morning, Dan was on a road-clearing mission en route to a potential laterite pit. As Dan recalled: "One of our officers on a helicopter spotted an artillery impact crater five clicks [kilometers] from the Di An Base Camp that appeared to have laterite. So, the first thing next morning, our sweep teams were sent on a road clearing detail."

Laterite was, after all, the best-available material for building roads, airfields, and other points of infrastructure. But because the nearest known laterite pit was *twenty* kilometers away in Long Binh, this new "potential" site could save time in gathering the coveted resource. The minesweeping detail consisted of two teams: one "comprised of SP5 Morris, who was in charge, along with SP4 Deal and SP4 Lester; and then the other team consisted of SP4 Baker and me." As Crowley recalled: "We were told the best route to get to the crater was to take Route 13, so we swept the road from Di An to Route 13. Then we had to clear an abandoned side trail that led in the direction of the crater."

Route 13 was the infamous "Thunder Road," where a number of high-profile battles would later occur between US and Communist forces. Several American bases were strung along the nodes of Route 13 and, during the early days of the war, its convoys were frequently targeted by Viet Cong ambush teams.

"At the entrance of the trail was a barricade of logs, brush… and an ominous large sign with skull and crossbones," a clear indication that mines had been detected ahead. Larry Blair, now barely six weeks into his command of Charlie Company, stood by the trail entrance with a platoon of infantry waiting for the minesweepers' arrival. "He had a couple of jeeps, a bulldozer, a bucket loader, and three dump trucks filled with dirt," Dan recalled. "We began to sweep the trail; and sure enough, we find our first mine of the day, and we blow it in place. We kept going… found two more, and blow them in place, too. The dump truck fills the holes as quickly as we blow the mines."

Dan and his teams worked their way up a gradually-rising slope, with the convoy close behind them. The minesweepers soon crested the top of a knoll, but as they began moving down the reverse slope (with the infantry on their flanks), the convoy stayed at the top of the knoll. About 150-300 yards from the convoy, Specialist Baker stopped Crowley, and shouted:

"Oh Shit! Look at that!"

Dan looked on in astonishment.

"There lies a menacingly large mine that had been buried in the ground for a very long time," he recalled. The rains had washed out the dirt around it, leaving the top of the mine visible with about one inch of ground clearance. But as Crowley and Baker went searching for a command wire, one of the convoy lieutenants, whom Crowley never identified by name, "was calling us all back to return to the top of the knoll. He was a little too late, though, because as soon as I spotted the mine, I started rigging a charge together with a piece of time fuse, blasting cap, and half-pound block of TNT. Everybody else had already headed up to the knoll."

Crowley prepped the charge, yelling "Fire in the Hole!" which Specialist Morris echoed to alert the convoy of the impending detonation.

But somehow, one of the bulldozers didn't hear the admonishment.

"I'm looking up at the top of the knoll," says Dan, "when I see the dozer start to move off the trail. It gets about five feet before there was a flash and a big bang."

The bulldozer had just run over an undetected mine.

"The dozer went up on one side and the driver is airborne. It crashes down hard, and about 16-18 GIs along with Captain Blair, who was standing nearby consulting a map, were catapulted off their feet…wounded by the shrapnel. The dozer driver landed hard and broke his shoulder."

About that same time, the Battalion Commander and Sergeant Major were circling overhead in their Huey helicopter. "They had been observing the operation…and began to circle around looking to land on the road to pick up the wounded."

But Dan still had to force-detonate the mine in front of him.

As Crowley yelled "Fire in the Hole!" again, the convoy lieutenant scrambled for the radio, and waved off the Huey just in time before Dan's charges detonated. As Dan recalled, the resulting crater was big enough to accommodate a Volkswagen Beetle—an impressive feat for an anti-personnel mine.

After the "All Clear" was given, the Huey returned to pick up the wounded. However, a dismayed staff sergeant exited the helicopter, asking the lieutenant to identify the culprit who had detonated the landmine him about the land "that nearly blew them to Kingdom Come."

Without batting an eye, the lieutenant pointed to Crowley. "I'm about to get a serious ass chewing," he said.

The indignant sergeant demanded to know why Crowley hadn't just dug up the mine instead of blowing it.

"Are you serious?" Crowley was livid.

This staff sergeant obviously had no idea how minesweepers worked.

Then came the round of non-sequitur questions. "He sees I have an M-79 and an M-16, and asks how did I come to have a M-16? Then, he sees that I'm wearing only a helmet liner with a camouflage cover, and no steel pot. He goes apeshit about this for a minute, and I gave him a bullshit story about it might fall off while working on a mine. I also showed him all the explosives in my demo bag along with twenty M-79 rounds, and a couple of grenades. His eyes got big and he began to step away. Nothing more was said. His attitude changed quickly, and he avoided me thereafter."

Perhaps this staff sergeant was afraid that Crowley intended to "frag" him, or even self-detonate.[10]

"With the bulldozer FUBAR, we returned to Di An."[11]

Meanwhile, Larry Blair grappled with leading an engineer company whose ranks were turning over almost daily. But with each incoming replacement, he took great strides to integrate them into the formation, and always led by example. Some commanders led from the rear; and they were content to manage their formations from the confines of their command vehicles.

Not Larry Blair.

No matter the mission, he always led from the front.

Sometimes, however, he got more than he bargained for.

---

[10] "Fragging" is the deliberate killing of military comrades (often unpopular leaders), typically carried out with fragmentation grenades, as they left no ballistic signature and could easily be attributed to enemy fire.

[11] "FUBAR" is a popular military acronym for "F*cked up beyond all recognition."

During one patrol, for example, he had a close call with a VC roadside bomber. "They were guerrilla fighters," he said, "so they would resort to ambushes…and we did the same to them. But if the battle wasn't going their way, they would disengage and slide back into the jungle." That being said, Larry acknowledged their skills as hit-and-run tacticians. "They were very good at it," he said, "lots of booby traps and mines…that sort of thing. Very innovative, too. They not only used captured mines from the French and Americans, but old Chinese mines left over from earlier wars. And when they couldn't get a landmine they wanted, they made their own."

Indeed, the VC would deconstruct other explosives and create their own mines.

"They made their own version of a claymore mine," Larry continued, "which was basically a pie tin…and they filled the front face with nails, bolts, rocks, and anything else they could make for shrapnel. And it was really deadly." His first encounter with the Communist-grade claymore occurred on the tail-end of a convoy ambush. The VC triggerman had detonated his bomb against a deuce-and-a-half truck, the blast from which had penetrated the engine block.

When Larry dismounted the convoy, he immediately went searching for a command wire. The bomber had to be nearby, and following the command wire from the blast site would lead them right to his redoubt. "Knowing that we would trace the wires back," said Larry, "the bomber had set *another* one," hoping to skewer any search patrol that came looking for him. "And so we went following the wires…and we find this little piece of red cloth hanging on the bush." Larry slowly parted the leaves on the bush…bringing him face-to-face with the enemy claymore.

"I about filled my drawers right then," he laughed.

But the humor was ironic.

He realized that the VC triggerman had "chickened out" and fled the scene, rather that detonate the secondary claymore intended for Larry's foot patrol. "We found the hole where he had been sitting, the firing device, and the batteries." Had the enemy triggerman not fled, he could have easily killed Larry and a few others by blasting the second claymore.

But these impromptu claymores were just the beginning. "When we started clearing jungles with bulldozers," Larry continued, "the VC would hang booby traps in the trees," which would subsequently fall onto the dozers, killing their operators. "So, we buttoned up our tanks to clear the initial squares out of the jungle," he said, "and then we'd finish the job with the bulldozers. When the mines would go off in the trees, they wouldn't hurt the tank. But they could make a bad day for the dozer operators."

Moreover, when the VC realized that American forces were using metal detectors for minesweeping, they started making landmine detonators from strips of bamboo. As Larry recalled: "It was two pieces of bamboo about eight inches long. And there's a rock on each end…wrapped with duct tape…to keep the bamboo apart. Then, right in the middle, there's two little pieces of tinfoil, like off of a gum wrapper, that are connected to the wire. When you squeeze those bamboo pieces together, you close the circuit," and the landmine would detonate.

Even the most seasoned minesweepers couldn't detect the bamboo explosives.

For anti-vehicle operations, the bamboo mines could disable a tank or APC, but not for long. At most, the mine could blow off a track, but it would yield no catastrophic kills. Undeterred, the VC started placing these mines offset, deliberately targeting the vehicle drivers. "They'd put the firing device where the tread would run over," said Larry, "and the mine itself would be three feet away, so it would go off right under the driver's compartment…or right under the belly of the tank."

And although the VC couldn't stand toe-to-toe against the 1st Engineers, they were quite adept at night infiltration. Aside from building and maintaining their tunnel networks, VC combat engineers were expected to penetrate American lines. One day, Larry witnessed a captured VC sapper give a demonstration of how they could penetrate an Allied perimeter. "With a pair of wire cutters, a rag, and a 20-pound satchel charge on his chest… he laid on his back and slithered underneath the [concertina] wire…and as he slithered under, he would clip his way through it.

He would use the rag to deaden the sound of the clipping. When you clip the wire, it makes a pop. But he would wrap that rag around the cutter, and you wouldn't even hear it."

Still, he marveled at how well his own troops thwarted the VC at every turn. Even during the most well-timed ambushes, the Viet Cong were never truly able to gain the upper hand. Moreover, despite the turbulent personnel system, the incoming troops seemed to be carrying on with near-seamless efficiency.

But the life of a company commander was not without its heartaches.

One of his worst memories involved the tragic death of a "short-timer"—a soldier with only a few weeks remaining in their deployment before going home. "There was an ambush patrol we mounted out of Di An one night," Larry recalled. "It was not too long after I had gotten there [to Charlie Company], and we were shorthanded because people were going home." Replacements were coming in, but there was always a lag between a departing soldier's exodus and his replacement's arrival. Because the ambush patrol was so short on personnel, Larry decided to go with them.

Suddenly, Larry was approached by a young specialist.

"Captain, I only got two weeks to go," he said. "And its SOP [Standard Operating Procedure] that when you have one month ago, you don't have to go out on any ambush patrols anymore."

Larry didn't know what to make of the remark.

"Well," said Larry, "I've read the SOP and I don't see that."

"It's customary," the soldier replied, "you know, I'm gonna go home in two weeks."

Larry sympathized with the young soldier. But they had a job to do.

"Listen," said Larry, "I understand that, and I would love to let you stay behind. But we are *really* shorthanded. I just don't have anybody else to go. You gotta go."

The young soldier acknowledged the order and, somewhat begrudgingly, draped a couple bandoliers of M60 ammo across his chest, assuming the role of an assistant gunner for the M60 machine gun.

The patrol might have been uneventful, had it not been for the weather.

"Just then, it started pouring down rain," Larry recalled.

"And we were crossing the dikes between rice paddies…we thought that there were just rice paddies on both sides. So, if you slipped off the dike…you'd fall into water up to your knees, and that was about it."

But little did they know that one dike had a 15-foot retention pond on its opposite side. "Well," Larry continued, "he slipped off that dyke, and with those two bandoliers of ammo, plus his web gear [load bearing belt and tactical suspenders], he went to the bottom." Larry frantically tried to save the sinking soldier, but to no avail.

"It was really a bad scene."

Some of the other soldiers were so traumatized by the drowning, that they couldn't carry on with the mission. They literally had to be taken away on the MEDEVAC helicopter. Charlie Company was finally able to retrieve the body of their drowned comrade the following morning. "That was a really bad chapter in my experience over there." A man with two weeks remaining on his deployment had fallen not to enemy fire, but the happenstances of Mother Nature.

Another "bad chapter" for Larry Blair was saying goodbye to his most seasoned junior officers. One such departing asset was Chuck Humphrey. Throughout the Battle of Bong Trang, Chuck's platoon had remained at Phu Loi in support of the LAM SON II pacification mission. And by the end of September, he too, was approaching his exodus date from Vietnam.

"I rotated home on September 29, 1966," said Chuck. "I was the last officer in the 1st Engineer Battalion (of those who deployed to the RVN in September 1965) to be allowed to 'go home.' The reason was that the Army decided to extend me in Vietnam for a couple of weeks, then release me from my 24-month active-duty requirement a month early. I was issued a DD-214 [official military discharge paperwork] at the Oakland Army Terminal and paid about $1,000 in unpaid vacation leave (in cash). I was away from the USA for 376 days, but my 24-month active-duty commitment ended after 23 months and 11 days. My date of release from active duty was September 30, 1966."

Meanwhile, Dan Crowley readied himself for another trip down Thunder Road. "What a ball-buster," he said, "lots of heavy lifting of logs and chain saw work. There were areas we had to cut back from the sides of Route 13 to make the road more visible from the air." Route 13 was, after all, "the major resupply route due north of Saigon to the rich rubber plantation areas of Lai Khe, Quan Loi, and Loc Ninh." The road was vital to the local economy; but north of Lai Khe the Viet Cong had been mining and patrolling Route 13 with virtual impunity. Division Headquarters was thus determined to break the VC's hold along the northern edges of the road.

For Dan Crowley, however, the biggest obstacle was simply felling the trees. "Often, the chain saws would fail because of the latex in the rubber trees," he said, "and we would have to dip the chains in diesel gas mix to keep them from seizing up." Most of the trees otherwise fell without incident, but the Viet Cong couldn't resist setting a booby trap or two. "In a couple of locations," Dan recalled, "the VC placed logs and brush piles in the road that concealed mines-you never knew what to expect." Whenever Charlie Company found a mine, Crowley and his team would force-detonate it—"and then use the logs to fill in the holes in the roadbed."

Dubbed Operation TULSA, this road improvement mission spanned from October 9-15, 1966. "The engineers were well aware of the magnitude of their job. There were many miles of roads to be cleared of mines and miles of badly-deteriorated surface to be repaired." Because the monsoon rains had been underway since early summer, however, many of the potholes and mine craters had become lakes unto themselves.

At around the same time, Dan noticed that the local VC agents had resorted to extortion and political assassinations. Some nights, he recalled, "the VC would set up operations for VC tax collectors," strong-arming villagers for money or supplies, "and

VC hit squads for any unfortunate government employee." Dan wryly recalled that: "If you weren't VC communist, you were immediately executed."

But the irony wasn't lost on Dan or his comrades.

Indeed, if the VC had to coerce or intimidate the South Vietnamese into supporting Hanoi's government, then their cause was not an appealing one.

Soon after Charlie Company began their work along Route 13, MACV opened the road to artillery and tracked vehicle convoys. The authorization was premature (the road wasn't truly ready, after all) but the Division-level mission requirements demanded it. As such, "it became increasingly apparent that the engineers' task of holding the road together was going to be an agonizing job." The Battalion Commander went as far as to say: "At times it looked as if the whole road would shortly sink into the swamp."

Dan Crowley and his fellow engineers continued placing timber treadways over the soft spots and, by October 11, it appeared that their section of Route 13 was suitable for convoy traffic. The first convoy was scheduled to pass through the following day, but on that night of October 11, the sky opened up for one final monsoon.

The felled timber once again lost its dry resiliency.

To make matters worse, Charlie Company had once again been converted to infantry. "This was the third time," Dan exclaimed, "we were assigned to secure an artillery position." Thus, Charlie Company had to clear the road of mines/craters by day, then return to guard the firebase by night. "It was wet, cold and the mosquitos were brutal."

In fact, that morning of October 12, the "tired and weary engineers faced a wet, sagging, Route 13…with the first resupply convoy only two hours away." But thanks to the frantic efforts of Charlie and Delta Company—"using timber, perforated steel planks, nails and shovels,"—the diehards saved the day, and the convoys passed without delay.

Throughout Charlie Company's mission along Route 13, Captain Blair stayed busy coordinating (and participating in) the various tasks that befell his engineers. As luck would have it, one mission involved using an unexpected "gift" from the Engineer Research & Development labs at Fort Belvoir—a Jeep-mounted mine detector. The scientists at Fort Belvoir had shipped the new mine detection vehicle into theater, and the 1st Engineer Battalion had been selected as the inaugural "beta tester."

As Larry described it, the mine detector was "something they had mounted on the front of an M38A1 Jeep [the WWII-era Willys variety]. It had a fiberglass head about eight feet wide, and mounted on like a snowplow to the front bumper of this Jeep."

Moreover, the vehicle was *remote-controlled*.

"You had to start the engine on the Jeep, put it in gear, and then they had a little control box, so you would follow in another vehicle…a hundred yards behind or a safe distance back. Using this control box, you could make the [radio-controlled] Jeep turn right or left, start or stop."

But, as Larry admitted, the contraption was "really crude."

In fact, testing and evaluating the radio-controlled Jeep seemed like an exercise in futility. It was a great concept, but the technology was years behind. "The steering was *very* spastic," he said. The slightest twitch of the controller would jerk the wheels into a hard turn. And a slight twitch in the other direction would overcompensate the turn.

"It was all over the place," he continued.

"The idea was that if it detected a mine, it would automatically disengage the clutch, and slam on the brake so the Jeep itself wouldn't run over the mine. Then you'd have to go up with a manual mine detector. You'd have to back the Jeep up, then find the mine and neutralize it…then move on with the mine detector." Theoretically, this radio-controlled Jeep could cut the minesweepers' mission time in half.

Larry was never fond of the machine, or its jerky handling, but the RC Jeep did make for some interesting color commentary. For example, during a test run along the highway, the driverless vehicle spooked some unsuspecting Vietnamese farmers. "There was a lot of wood stacked up on both sides of the road," he recalled. "And Vietnamese farmers would come out there with their ox carts and cut up all the wood that we had knocked down. Then they'd haul it back into their villages," to use as kindling. But aside from their timber scavenging, the farmers would never go on the road. Most of them didn't want to risk hitting a random VC mine. "Still, we would see them out there every day chopping up the trees and hauling them out in their oxcarts. One day, we decided we wanted to sweep that road with this Jeep-mounted mine detector."

One of the problems with this remote-operated Jeep, however, was the time lag for the signal to disengage the clutch. "It wouldn't do it quick enough," Larry said. Thus, to gauge exactly when the clutch disengaged (and subsequently stall the engine) Larry would turn on the windshield wipers. "If you could see the windshield wipers going back and forth, it meant the engine was still running."

Driving down the road, however, this erratic, driverless Jeep ambled by a group of Vietnamese farmers. "They see this Jeep coming down the road being chased by another Jeep with three or four guys in it, and there's nobody in the first Jeep. And it's going all over the road, back and forth, up and down…the windshield wipers are going and there's no rain." The Vietnamese were, by tradition, a very superstitious people…with long-standing beliefs in the supernatural.

"And they're thinking this Jeep was being driven by a crazy ghost!" he laughed. The terrified farmers whipped up their oxen and beat a hasty retreat…never to return. "Mercifully," Larry continued, "one day that Jeep hit its own mine and blew itself up. So, we boxed up all the parts and sent them back to Fort Belvoir. That's the last we heard of the Jeep mine detectors. Unfortunately, I never got a picture of that stupid machine. But there is one in the museum at Fort Leonard Wood," he laughed.

Following Operation TULSA, Charlie Company returned to Di An. For Dan Crowley's 1st Platoon, it was back to the technical work of base construction—pouring concrete pads for the camp's new tropical billets. "The billets had metal roofs, screen sides, and even screen doors," said Dan. "They had sandbags up to five feet tall around the building. Our Company finally got to move out of our tents and into one of the last billets to be completed. But we only got to occupy them a short time before we were rotated back to Lai Khe." As Dan summarized it, returning to Lai Khe was like going "back to the ghetto."

Yet somehow, Charlie Company was the *last* to know about their impending move to Lai Khe. Indeed, the local *femme fatales* knew about it before the GIs did. One morning, Dan had been doing road repairs along Route 13 outside of Di An when he was approached by two call girls.

"GI! You come to Lai Khe next week!" they said. "We have #1 good time!"

Dan laughed off the remark.

But one week later, he was astounded by the news. "By golly, they were right! In a week's time, Charlie Company was headed to Lai Khe. They knew before we did."

Going back to Lai Khe was no proverbial "picnic," but working along Route 13 (even near Di An) hadn't been pleasant either. "There was lots of civilian traffic," Dan recalled—"bicycles, motor bikes, ox carts, and 3-wheel scooters designed for two people, but with *five* or *six* on board." It seemed that the Vietnamese prioritized efficiency over vehicular safety. "There were people on foot," he continued, "along with buses that looked like zoos on wheels." Indeed, many of these buses had been packed with both people *and* livestock—including cages of ducks, geese, chickens, and pigs. Normally, Dan wouldn't have minded the high volume of civilian traffic. But considering that the enemy lurked in and amongst the population, every passing car or overcrowded scooter might be carrying a suicide bomber.

Back at Lai Khe, Dan Crowley was surprised to see how little had changed over the past seven months. "We still had holes in our tents, no real billets, or new outhouses," he said. "The monkeys were still nasty, smart, bold, and stole anything that wasn't nailed down."

But there was a light at the end of his tunnel.

"My time was getting short," he said, "with less than 60 days left in country."

All he had to do was stay alive through the end of December. But there would be *no* slowdown in his mission tempo.

"We had been at Lai Khe only a couple of days when we got the word 'Saddle Up,'" he recalled. "We were going to cut an LZ using the C-47 Chinook, which meant we had to use a rope ladder insertion." Operation BATTLE CREEK (later renamed "ATTLEBORO") had begun.

It would be Crowley's last hurrah in the Republic of Vietnam.

CHAPTER 7
THE DEMO EXPRESS

At first, it seemed that November would be a quiet month—at least for Charlie Company. Vertical and horizontal construction, while not easy work, was still a welcomed break from the perils of combat. But, late one evening, the sustained roar of Allied aircraft shattered the nighttime calm—alerting the men of Charlie Company that something "out of the ordinary" was happening. As it turned out, elements of the 196th Light Infantry Brigade and the 25th Infantry Division were engaged in heavy combat with the 273d VC Regiment, the 9th VC Division, and the 101st Division of the North Vietnamese Army, all of whom were operating in the Tay Ninh Province. The 1st Infantry Division (and, by extension, the 1st Engineer Battalion) would soon be thrust into the fury of Operation BATTLE CREEK, which would later be absorbed into the larger Operation ATTLEBORO.

The 1st Engineers were alerted for duty on November 4. At one minute past midnight the following morning, Charlie Company (along with Alpha and Bravo) began moving south to Suoi Da, where they would cut the LZs needed for heliborne insertions. As Dan Crowley recalled: "1st Platoon was assigned to go in first. Charlie Company loaded onto one of the two Chinooks." The trailing Chinook carried a sling-loaded pallet—"packed with 2,500 pounds of demo, chain saws, gas, water and assorted hand tools."

Dan was continually amazed by what these old utility choppers could carry.

Crowley himself, meanwhile, carried a basic load of ammunition—"plus a 20-pound satchel filled with high-explosive charges along with a coil of detonation cord." The flight to Suoi Da was uneventful, but dismounting the rickety old Chinook seemed like a near-death experience. "Climbing down through a hatch in the floor of a Chinook by hanging onto a rope ladder," said Dan, "was a real trip -with a very tight pucker factor."

All he wanted to do was get his feet *flat* on the ground.

Luckily, they hadn't landed under enemy fire.

"Right before our arrival," Dan continued, "the LZ had been bombed and prepped by our artillery and strafed by helicopter gunships. We were tasked to make an opening in the jungle for the other Chinook to get in low enough to drop us the sling load."

Dan's team had no contact with a security element.

An infantry unit would typically provide overwatch while the engineers worked.

Had a security team even been posted? They didn't know— "but we got to work anyway," he said—the second Chinook needed a place to drop its supplies…and quickly.

"We ran detonation cord through the jungle, and then tied the charges placing them next to trees, bamboo clumps, and termite mounds that were six feet tall," he continued. "We used about 3,000 pounds of demo on our first charge." After setting the charges, Dan yelled out the customary call:

"Fire in the Hole!"

Instinctively, every demolition man knew that "what goes up, must come down." That's why Dan Crowley always gave himself enough distance and defilade before pulling the trigger. But with today's demolition, there was a slight problem: "Ants!" he cried. "They were very large with menacing pinchers and they lived and nested in the trees. After blasting the surrounding trees, the debris would fall down with lots of pissed-off ants." But these ants were unlike any he had seen in the US. Indeed, they weren't truly fire ants *or* the garden-variety "army ants." These ants had pincers that felt like vise grips. Dan recalled that, when clearing twigs and branches from the landing zone, the ants would jump on their clothes, and bite down with all their might. "Those demon bastard ants were everywhere," he said, "and when their pinchers clamped down, you had to really pull *hard* to get them off. Not Fun!"

Later that afternoon, Crowley's demolition team ventured deep into the jungle where there had been a major battle that morning. The engineers had been ordered to pick up a cache of captured weapons, and take it back to the LZ they had just cleared that morning. "We were to load it up on the helicopters," said Dan, "but before doing so, the 'Brass' [high-ranking officers] had to have their photo op with all the captured loot." Whatever the helicopter couldn't carry, the demo team blew in place. "It all had to go," said Dan. "As they say: 'Fire in the Hole!'"

As it turned out, that "major battle" had been between the Viet Cong and the 1-28 Infantry battalion. The latter had penetrated deep into the jungle and discovered the "largest cache of enemy ordnance ever captured up to that time." The total yield included some "25,000 grenades…55 rifles and automatic weapons, 752 Bangalore torpedoes, 11 water mines, 505 claymores, and other war materials." Other elements of Charlie Company, meanwhile, destroyed the "bunkers, a machine shop, and a VC hospital in the same area"—all while under repeated sniper fire.

By nightfall, "most of Charlie Company, including our infantry security, had been air lifted out, leaving twenty to thirty of us still on the ground."

No worries, thought Dan.

Two more round trips, and all personnel would be off the LZ.

But after a while, Dan started to worry. "It was dark and the choppers had stopped." Several minutes passed…and yet there was no telltale sound of a chopper blade on the horizon. "We had no perimeter or foxholes dug."

Had they been forgotten? Were they now stranded at the LZ?

Being a E-4 Specialist, Dan admitted: "I'm the last to know anything…and I'm beginning to worry."

But suddenly, a voice came over the radio.

"Night Movement."

All right. But where?

An infantry battalion's perimeter—"four hours away, through triple-canopy jungle, at night. *Are they nuts?!*"

Dan was livid.

But none were more upset than Larry Blair.

Larry, along with Dan Crowley, had been among the few dozen troops left behind at the LZ. As the company commander, he was the ranking officer amongst the tired GIs who lingered on the Suoi Da LZ. And now, he had the unenviable task of convincing his men that a *four-hour*, nighttime patrol (through enemy territory) was a feasible way to link up with 1-28 Infantry.

"Off we go with Captain Larry Blair in the lead," said Dan, "with the only compass and red lens flashlight in the whole detail. It's pitch black and we are strung out holding onto the man in front of you by his web gear." Dan grabbed hold of the Vietnamese interpreter in front of him—"and we are moving like a herd of turtles." Somewhere in the distance, Dan heard the faint rumble of artillery fire.

"They were firing parachute flares for us."

Indeed, a local gun battery was firing illumination rounds. These falling flares would light up the area long enough for a foot patrol to find its way in the dark. "It looked like a kaleidoscope of flickering lights under the thick jungle canopy," Dan recalled, "and then it would go pitch black again."

But friendly flares weren't the only things he'd see that night.

"Soon after we started our movement," Dan continued, "we began to get 60mm incoming mortar rounds." It seemed that the illumination rounds were likewise helping the Viet Cong; because they were now launching mortars at the slow-rolling American patrol. But despite the intermittent flashes of light, the cover of darkness still favored the Americans. The VC couldn't accurately target their mortar fire. "They were hitting to the rear, then to the right, then to the left of us," Dan recalled. "A lot of stuff would hit high in the canopies, ricochet off of branches, and come crashing down around us."

But the VC were entitled to at least one "lucky punch."

"All of a sudden," said Dan, "something came crashing down and the interpreter is cold-cocked. He goes limp. Out cold!" Since the round didn't explode on impact, Crowley assumed it had been a flare casing instead of a mortar round. After all, gravity worked the same for flare casings as it did for mortars.

But, years later, at a 1st Engineer Reunion, Crowley discovered the truth.

Speaking with Larry Blair, Crowley mention the "flare casing" when Blair abruptly stopped him. "Those things crashing through the canopy were 60mm duds," said Larry. "And if that round had detonated –you and about six others would have been history."

Dan was speechless.

"Death had *only* been an arm's length away."

They thrust the still-unconscious interpreter off the jungle floor and kept moving. "About a half hour later," Dan continued, "we reached the perimeter of the 1-28 Infantry. To this day, I believe that was one heck of a piece of navigation by Captain Blair." Dan was convinced that the nighttime maneuver itself warranted Blair a medal.

As the weary patrol hobbled into the perimeter, it started to rain. It was the perfect end to a horrible night—"We were cold, beat, tired and sore all over, and the merciless mosquitos were hungry." Now within the confines of 1-28, Charlie Company once again found themselves as impromptu infantrymen.

"We were to be used as a 'fill-in' for the perimeter," said Dan.

Charlie Company would provide security by night (using foxholes dug by the frontline grunts) and perform their regular engineering duties by day. "Sergeant Jauron and I were placed in a well-used foxhole that had been utilized in the previous morning's battle," Dan continued. "By flare light we could see remnants of M-60 brass, links, and used field dressing packs. Under a sandbag, I found a jammed .45 caliber pistol, which I cleared and handed to Sargent Jauron." Crowley was already carrying three other weapons; he didn't need a fourth.

"The next day, we were moved to an artillery fire support base to guard it." As they settled into another round of perimeter security, Dan realized that this was the *fourth* time that his unit had been converted to infantry during the past 11 months.

But this time, no one was complaining.

As Crowley described it, guarding the new firebase was somewhat of a glamour detail. "Hang out in the perimeter, sleep all day." At one point, he half-jokingly suggested changing his MOS to Infantry. "It beats carrying bridge panels, running a chain saw, or sweeping a road," he said.

But life in the Infantry had its own perils.

One night, for example, Crowley and two others were sent to an "Observation Post" (OP) to act as the nighttime "eyes, ears, and early warning system" for the whole firebase. If the VC were lurking beyond the perimeter, the OP would radio the Command Post (CP), coordinating an artillery strike from the firebase while the OP troops held the enemy with small arms fire.

Specialist Deal, who had been on light duty after sustaining a rat bite, was put in charge of the OP, accompanied by Crowley and Specialist White (the platoon's radioman). "We went well-armed," said Dan, "with M-14s and one M-79 with canister rounds; plus, I had my Claymore." It had been raining for most of the day and it was still raining by the time Crowley's team left at nightfall. "We were instructed to go 100-200 yards in front of the base perimeter," Crowley continued. "Our job was to be on alert for any enemy activity, and report it back to the CP."

Because of the rain, they slogged through ankle-deep water until they found a small mound—the only piece of relatively dry land in the vicinity. Crowley, Deal, and White were relieved to find a bit of "high ground;" but as Crowley admitted, they were sharing their new redoubt with "lots of desperate critters seeking a high spot, too." Dan described their wetland outpost as "eerie and scary," but they occupied their OP and stayed quiet.

"Just before dawn," he said, "we got the word to come in."

For sure, the trio was ready to come back.

They had been lying in wait for hours…and nothing had happened.

But all things considered, they had been lucky.

Some prior OPs had been overrun; some had even become victims of fratricide.

Crowley and Deal threw a couple of grenades out into the darkness and "hauled our soggy asses back to the perimeter." As Dan recalled, they didn't have a reason to blow anything up. "But just in case some VC were trying to sneak up on us," he said, "we took *no* chances."

Better safe than sorry.

But some wondered if the sudden grenade explosions might actually *draw* fire from the VC.

"After we were relieved of our infantry duty, we went back to work sweeping roads," and repairing the civil infrastructure—roadways, bridges, etc. "As part of Operation Attleboro," Dan continued, "we as engineers of Charlie Company had cut a landing zone; repaired roads; done a night movement through the jungle on foot; were used as infantry; and now we were assigned to cut *another* LZ using the Chinook once again."

Just as before, Dan Crowley climbed aboard one Chinook, while watching his gear—"chain saws, gas, and lots of explosives"—sling-loaded onto another Chinook. "We were ready to use the rope ladder again," he said, "but we got lucky." Indeed, this time, rather than descent amongst the treetops or natural clearings, a misplaced Allied bomb had created its own clearing in the jungle canopy a few days earlier. "The Chinook could hover over it while we jumped off the rear cargo ramp. The LZ we were about to cut was on top of a small mountain called Nui-Ong…and we were told the VC owned the bottom half - Great!"

Just as it had been at Suoi Da, the demo team had *no* contact with the infantry who were supposed to be providing security. "But we got to work laying out our line charges anyway." Dan's team leader, Sergeant Jauron had a special "formula" for blowing down trees.

"*Give it an extra block of TNT.*"

Dan happily obliged.

He set off the first charge, igniting 1,500 pounds of explosives. But as the dust, ants, and debris settled down, Dan's exuberance quickly turned to horror.

"I saw a mangled GI standing in the blast zone."

Crowley was mortified.

"*Oh shit!* Where the hell did *he* come from?!"

Dan couldn't figure it out. The demo men were supposed to be the only ones on the mountainside. Who was this random GI? And where had he come from? Neither Crowley nor Jauron recognized him. He wasn't a Charlie Company soldier.

"We called for a medic," said Dan. "This poor guy was hurt big time. Later, we found out that he was part of our infantry security. They had climbed up the mountain, but they were late in arriving. He had been sent up ahead by his platoon leader with a stringer of empty canteens to fetch water; and he happened to walk onto the charge just as it went off." Crowley put him on the first MEDEVAC chopper out. "I truly hoped he made it."

Meanwhile, Larry Blair took the rest of Charlie Company on a mission to the Special Forces Camp at Minh Tanh. Manning a weekly convoy from Di An to Bien Hoa, the engineers began loading their equipment onto C-130s for the flight into Minh Tanh. "We had taken all the stacks, headache boards off our trucks, graders, and loaders…so they could fit them into the C-130," said Larry. "We couldn't take our dozers, because they were too heavy for a C-130. So, we borrowed a bulldozer from the Australian Engineers, who had a camp right near Di An."

The Aussies agreed, but under one condition.

"If you take our dozer, you gotta take our operators."

"Hey, that's fine," said Larry.

"So, along came these two cocky Australian guys," he chuckled. "They were great guys, though…good operators."

And since the war began, they had proven themselves to be reliable allies.

"So, we flew into this Special Forces camp," Larry continued, "and because the area all around the camp wasn't secure, the C-130s would have to go in above 1,500 feet, take a steep dive, and stick it right on the runway,"—essentially making a short-contact landing. This would, theoretically, minimize the plane's exposure to enemy fire. "Flying up there in those C-130s was an interesting experience for everybody. We had a dump truck in each C-130…new road graders, trailers, and a couple of bucket loaders. All of those were flown in C-130s; they'd latch them down." But as the plane went into its steep dives, Larry would get a white-knuckle grip on his seat in the cargo bay…hoping the dump truck didn't break loose.

Upon landing in Minh Tanh, Charlie Company opened the nearby laterite pit, and began resurfacing the runway. At this point, the Special Forces camp had little more than a dirt runway. Blair's mission, therefore, was to lengthen, resurface, and reinforce the airstrip. For much of November-December 1966, Charlie Company worked on the "Little LaGuardia" airfield at Minh Tanh. "The runway of 4,500 feet was rebuilt," said Larry, "three taxiways were expanded and the aircraft turnaround was enlarged." On the western side of the airfield, Charlie Company cut down nearly 4,000 rubber trees, making way for a helipad and an artillery firebase. During the day, the Special Forces unit provided overwatching security for the engineers. At night, the engineers returned the favor—providing perimeter security for the camp.

All told, Operation BATTLE CREEK - ATTLEBORO had been another resounding success for Allied forces. In fact, American forces had won every battle since the "combat mission" began in Vietnam the previous year. Yet, strategic progress against the NVA and Viet Cong seemed difficult to quantify. The enemy would be defeated in a particular sector, the area would be deemed "secure," yet the ARVN couldn't seem to keep the VC away for too long. Indeed, new enemy cadres would soon repopulate the previously-cleared sectors. Such was the case after Operation ATTLEBORO in the Tay Ninh Province. Despite the battlefield victory, Allied forces did not eradicate the VC's political domination in the province. The VC simply retreated to their cross-border sanctuaries in Cambodia, and returned to Tay Ninh after the Americans withdrew.

Still, the immediate results of ATTLEBORO were hard to ignore. US intelligence estimated more than 1,000 North Vietnamese losses. Allied casualties totaled 155 killed and 494 wounded. However, the most significant result of ATTLEBORO was the damage it inflicted on the enemy's supply system. American forces captured several hundred small arms and mines; several thousand grenades; 1,135 pounds of explosives; and more than 1,000 tons of foodstuffs. At least in the short term, ATTELBORO had been a critical setback for the Viet Cong and NVA.

Back at Lai Khe, Dan Crowley settled into the final weeks of his tour in Vietnam. He was now a critical "short-timer." But even short-timers weren't beyond the reach of Mother Nature or the fickle hand of fate. Dan recalled that: "late one evening in early December, I was in a poker game along with our medic. I hadn't been feeling too good the last few days, and I had gone to sick call, but the doctor said: 'Nothing wrong with you. Return to duty.' I was playing my hand, and the next thing I know…I was on the floor shaking and sweaty, throwing my guts up."

It seemed that the doctor had made a critical misdiagnosis.

"They put me on the next Caribou out of Lai Khe headed to the Long Binh hospital [93rd Evacuation Hospital]. I had a rash, high fever, parasites, and was hospitalized for four days." On his last day there, Crowley was standing in line at the dispensary when, suddenly, another soldier bumped into him. "He was trying to talk to me. But he couldn't—his jaw was wired shut and his face was a mess."

Dan didn't recognize him, but this strange GI seemed to know Dan.

"He's using sign language trying to ask me something. I'm thinking he's a nut case until he grabs my dog tag and he shows me his."

To Dan's astonishment, the wire-jawed soldier was one of his childhood friends.

"It's Pat Graham! He and I were in Cub Scouts together."

Ten years earlier, they had been in the same Cub Scout pack in Northern Michigan. "What a small world," Dan marveled. "The conversation was short because of his jaw injury," Dan admitted, but it was nice to see an old friend, especially 8,000 miles from home.

"After my hospital stay, I returned to Lai Khe," he continued. But by now, Dan Crowley was just a few short weeks from the end of his tour. His year in Vietnam was coming to an end. As such, he found himself relegated to "light duty" while waiting for his flight back to the "World." Lighter duties typically befell the so-called "short timers," but their discretionary time often led to mischief. Decompressing from the horrors of combat, however, one could hardly blame them. In fact, Dan recalled two of his platoon mates raided a supply depot on a "midnight requisition" and came back with a bounty of jungle boots; fatigues; cans of salmon; ham; grape and apple juice. "We felt like we had hit the jackpot and partied hard."

With less than a week remaining on his tour, Dan Crowley was removed from field duty, but "old Murphy," as he called it, was "about to piss on my parade once again." He rode in a convoy down to Di Ain to be out-processed and, fortunately, the "convoy ride was uneventful." Once at Di An, however, the youthful impulses of a short-time GI were about to take over. "We were billeted close to the EM club," said Dan, "and one of my short-time pals had a bottle of Jack!"

Not one to pass up a victory drink, Dan Crowley enjoyed some of the fine spirits.

But perhaps he enjoyed it a bit too much.

"I do remember making it back to my bunk," he recalled, "but nothing after that." A few of his buddies tried to wake him in time for his flight, but they couldn't rouse him from his sleep.

Panicking, his friends alerted the medics.

Dan woke up back in the Long Binh 93rd Evacuation Hospital. "I missed my flight out of Nam on December 22," he recalled sheepishly. "I could have been home for Christmas. Instead, I ended up watching Bob Hope's Christmas Show with Phyllis Diller and Joy Hetherington. At least I had a good laugh. By New Years, I'm back in South Florida [on post-deployment leave] where I cannot buy a drink or vote. I'm only 20 years old and this is 1967.

"One of the first things I did when I arrived home," he

continued, "was buy a 1965 [Mercury] Comet Cyclone—4-speed, bucket seats, 289 horsepower, and 4-barrel carb. I spent my leave time hanging out at the beach, Snook fishing, diving; and at night, I checked out the bars and girls. I had just turned 21 after two weeks of being home."

Dan was out of Vietnam, but his time in the Army wasn't done yet.

"When my leave was up, I was headed to Fort Carson, Colorado - home of the 5th Infantry Division and 7th Engineer Battalion. I have a year to go on my enlistment and Murphy's Law is about to kick my ass again. I'm stopped at the main gate in civilian clothes with no post sticker on my Comet." The "post sticker" referred to an Army-issued windshield decal, indicating that the vehicle belonged to a soldier, and that the vehicle was authorized to be on the installation. "The MP tells me I need to be in uniform when I report to Company D, and I have two days to get a post sticker for my car."

No problem, Dan thought. It seemed like a reasonable task.

"I decided to cruise around the post before signing in," he said, "and I found an out-of-the-way place [still on Fort Carson] to change into my uniform." After changing clothes, however, Dan couldn't help but notice that the road was wide and featured "burn-out marks like in drag racing." Sure enough, this strip of road was notorious for lead-footed GIs. "Next thing I know," he continued, "a Camaro SS pulls up and asked if I wanted to run."

Certain that his V8 Comet was up to the task, Crowley accepted the challenge.

"You bet; let's go."

And off they went.

"Can you guess what happens next?" Dan asked rhetorically. "The MPs had been watching this particular road for drag racers and we got nailed. They write us up and we got delinquent reports. But it doesn't end there—I'm about to get the ass-chewing of my life by the biggest First Sergeant I'd ever met."

Still fuming from his close encounter with the MPs, Dan walked into his new company's orderly room…right past a large sign that said 'Knock Before Entering.' The big, hulking First

Sergeant, furious that anyone could miss the posted sign, erupted, shouting:

"How dare you walk into my orderly room without knocking?!"

"Boy," said Dan, "was he hopping mad."

Dan's heels were locked at a rigid state of attention. The First Sergeant was just beginning to run out of steam when, suddenly, the phone rang. "The clerk answered it and looks at me, then hands the phone to the First Sergeant as he took a short breath from ripping me a new one."

In an instant, Dan knew who was on the other end of that phone.

More to the point, he knew what was coming.

He was about to endure the second round of an "ass chewing."

Still on the phone, the First Sergeant said: "Yeah, he just came through the door." And when he hung up, the tirade against Crowley continued. "I didn't think there was any ass left to chew," Dan recalled, "but he had no trouble finding it." Luckily for Dan, he hadn't yet signed in when the MPs issued his delinquent report. Thus, he was still technically "on leave," so his delinquent report didn't count against D Company.

But while Dan Crowley readjusted to life in the "World," Larry Blair continued leading Charlie Company in the fight against North Vietnam. He spent Christmas at the Minh Tanh Special Forces Camp, finishing their construction work at "Little LaGuardia," before being re-directed onto Operations CEDAR FALLS and JUNCTION CITY in 1967. Both operations saw Charlie Company plunged into "deep jungle work," said Larry—"cutting landing zones, clearing massive jungle areas, minesweeping, and demolition work." Road work and daily mine sweeps constituted the bulk of Charlie Company's duty during JUNCTION CITY. On March 7, 1967, Charlie Company returned to Lai Khe, beginning construction on a bypass road around the camp.

It was the last construction project Larry would oversee as the company commander.

Now, Larry, too, was approaching his one-year term of service in Vietnam.

He relinquished command of Charlie Company later that spring, and redeployed to the United States with orders to the Engineer Officers Advanced Course (EOAC). An extension of the Officer Basic Course, EOAC was a nine-month residency course wherein young captains learned the fundamentals of higher-echelon staff work and company command. Of course, for Larry Blair and many of his returning contemporaries, the course was little more than a "rubber stamp" needed to fulfill the metrics of the officer education system. After all, Larry had already served on battalion staff, and led a company *in combat*.

There was little (if anything) that the course could teach him.

But Larry was eager to share the lessons he had learned in Vietnam. For if one of his classmates was headed overseas, the imparted knowledge just might save his life.

In July 1967, Dan Crowley earned his promotion to Sergeant. But he had no intention of re-enlisting. "Three years was enough for me;" and he eagerly awaited his forthcoming discharge. But as Dan prepared to leave the military, the events in Vietnam were about to take a curious turn.

Still, throughout 1967, it appeared that Allied forces were making strides against the Communists. That year, "some two-thirds of the [South Vietnamese] hamlets were judged secure and under the control of the central government." Moreover, US forces had killed nearly 81,000 Viet Cong and NVA. If these trends continued, General Westmoreland predicted an orderly withdrawal beginning as early as 1970.

However, in January 1968, the Tet Offensive drastically changed the course of American intervention in Vietnam. By then, Hanoi realized that the NVA stood no chance of defeating US forces in open combat.

The Communists, therefore, settled on a different approach.

If they couldn't defeat Americans on the battlefield, they would try to undermine the ongoing narrative of "rural pacification" and bring the US to the negotiating table. Thus, under the cover of a "ceasefire" during Tet (the Vietnamese New Year), Hanoi directed the Viet Cong, supported by select NVA units, to launch a massive, simultaneous attack on several key US and ARVN installations… as well as urban centers and provincial capitals throughout the country. It was a bold move and the Communists weren't even sure that it would work. But if it stood any chance to undermine South Vietnam's credibility, and shake America's confidence in the war effort, it was worth the risk.

Although the US and South Vietnamese forces effectively crushed the Viet Cong uprising, the American media painted a very different picture of the Tet Offensive. Television broadcasts showed frightening images of the Viet Cong storming the American Embassy in Saigon and bloody fighting in the streets of Hue and Khe Sanh. Taken together, these haunting images led Walter Cronkite, America's most trusted news anchor, to declare that the war was now unwinnable.

All at once, President Johnson and Secretary McNamara lost their credibility as public opinion turned savagely against the war. Scrambling to save his administration's dignity, Johnson announced on March 31, 1968, that he would not seek re-election and would instead devote his attention to ending the war in Vietnam. Although Tet was a resounding failure for the North Vietnamese, it gave the Hanoi bureaucrats exactly what they wanted—a crisis of confidence within the American war effort.

Dan Crowley, meanwhile, was suffering from the downstream effects of the Communists' PR disaster. His official discharge date was supposed to be January 30, 1968—the same day the Tet Offensive began. Due to the time zone difference (and delay of real-time information), he might have been able to clear Fort Carson without incident. "But if you remember, I spent two days in the Navy Brig back in Oakland on my way to Vietnam. That was December 25, 1965; and the Army never forgets." Indeed, the Army was making him stay at Fort Carson for an additional 48 hours because "I had two days of 'bad time' to make up."

During the interim, Fort Carson went into "lockdown," and all pending discharges were put on hold. The Adjutant's Office told him that until Tet could be resolved, the Army didn't want to lose any critical personnel. And, as a combat engineer, Sergeant Dan Crowley was designated as a "critical MOS." However, the panic surrounding Tet soon subsided, and the Army turned him loose. "Just like that, my Army career ends and my life begins."

But although Tet's shockwaves had died down in the Pentagon, the long-term damage had been done. Amidst this public backlash, Dan Crowley and Larry Blair noticed that the American people were growing increasingly hostile towards the military. Indeed, soldiers who deployed to Vietnam with the Class of 1965 were returning to a very different society than the one they had left. The inspired patriotism of John F. Kennedy's "New Frontier" had disappeared in the social unrest and political turbulence of a "Great Society" that hardly lived up to its name. Even worse, some Americans were blaming the military simply for its involvement in the war. Dan and Larry could hardly believe it—US servicemen who had once been heralded as "heroes" were now being protested, spat upon, and called "baby killers."

But for many Vietnam veterans, the worst was yet to come.

EPILOGUE
REQUIEMS & AFTERSHOCKS

The 1st Engineer Battalion remained in Vietnam until 1970, rotating personnel every twelve months (unless killed or medically evacuated). The battalion redeployed to Fort Riley with the rest of the 1st Infantry Division, whereupon the 1st Engineers reassumed their "mechanized" posture. By the time Charlie Company returned to Fort Riley, however, President Johnson's turnabout on Vietnam had morphed into Richard Nixon's policy of "Vietnamization"—a redeployment of American forces while training the South Vietnamese to take the lead in combat operations.

Vietnamization was essentially a three-step process: (1) Increase the ARVN's combat and logistical capabilities; (2) Systematically return the Corps Tactical Zones to South Vietnamese control; and (3) Withdraw American ground forces. By order of President Nixon, MACV's posture was to be one of "defensive reaction." In other words, American troops were expected to let the enemy come to *them*.

But men like Dan Crowley, Larry Blair, Chuck Humphrey, and Jay Franz doubted the feasibility of a *defensive* posture. By 1967, the Pentagon's "top brass" knew that neither American forces *nor* the NVA could truly "win" the war in Vietnam. By most accounts, it seemed that Vietnam was devolving into a stalemate, much like Korea. Allied forces had gone to engaging the enemy in smaller skirmishes. In fact, by 1970, many thought that the war in Vietnam would end in a manner similar to Korea—with both sides accepting a cease-fire and settling for the *status quo antebellum*.

But Vietnam was no Korea.

The socio-political dynamics at home and abroad were much different now. And the resulting unrest had eroded much of the US Army's vitality. Some protestors began burning their draft cards, and the entry standards for new draftees fell progressively lower as the Army scrambled for bodies.

In fact, one brigade commander in Vietnam opined that the situation would get worse.

"The problem of motivation will grow as the level of combat winds down," he said. "North Vietnamese doctrine has always stressed that a *real* victory is a victory over the opponent's spirit." Indeed, body counts were meaningless to North Vietnam if America's discipline and morale were wavering.

Despite the political victory of Vietnamization and Nixon's promise of a "Peace with Honor," public opinion was still savagely against the war. Returning troops were still being targeted by angry protestors and, to make matters worse, a new protest group of disillusioned veterans had emerged from the unrest: Vietnam Veterans Against the War (VVAW). Staging rallies, organizing marches, and even testifying on Capitol Hill, VVAW's members included John Kerry (future Democratic Senator and 2004 presidential candidate) and Ron Kovic (*Born on the Fourth of July*).

On college campuses, student protests had begun targeting the schools' ROTC programs. At the height of Vietnam, these protests had become so intense that several ROTC units began ordering their cadets *not* to wear their uniforms on campus.

Amidst the public backlash, however, Hanoi seemed poised to pursue peace. Although the Tet Offensive had created the "crisis of confidence" they had hoped for, the tactical performance of the NVA and Viet Cong had been less than stellar. Tet had rendered the Viet Cong ineffective; the US invasion of Cambodia had disrupted the NVA's regional supply nodes; the increased bombing campaigns against North Vietnam (Operations LINEBACKER I and II) wrought havoc on Communist morale; and the NVA's

Easter Offensive of 1972 had been a disastrous failure.

Thus, it came as no surprise when North Vietnam came to the negotiating table for the Paris Peace Accords in 1973. The resulting truce called for an immediate cease-fire, a full-withdrawal of American forces, and a permanent demarcation between North and South Vietnam. The North Vietnamese, however, had no intention of abiding by the agreement.

After Hanoi was certain that the US would not intervene, the NVA invaded South Vietnam in the Spring of 1975. In response, ARVN troops mounted an unsuccessful defense before retreating through Hue, Danang, and finally, Da Lat. With the North Vietnamese closing in on Saigon, ARVN troops made their final stand at the Battle of Xuan Loc. As NVA tanks rolled into Saigon on the morning of April 30, 1975, the South Vietnamese government finally surrendered.

Watching these events unfold from the US, the veterans of Charlie Company were incredulous. In Vietnam, the US had won every battle, yet somehow managed to lose the war. After nearly a decade of misguided war policies (and over 58,000 combat deaths) America had abandoned an ally whom she had promised to defend. Watching the last helicopter evacuate from the rooftop of the American Embassy in Saigon, it seemed as though the entire war—the bloodshed, the sacrifice, and the untold cost in human suffering—had been an exercise in futility.

Meanwhile, Crowley, Humphrey, Blair, and Franz all tried to find a new sense of "normal" in their lives. They had gone to war with a firm belief in their mission…backed by the full faith and credit of the American people…yet came home to a hostile public that resented their very existence.

Each man eventually returned to civilian life. And each man made his own journey to heal from the physical and emotional wounds he carried with him.

After Dan Crowley returned to Florida in 1968, he lived a nomadic lifestyle for the next thirty years. Maybe it was his pent-up wanderlust…or a subconscious return to the habits of his itinerant childhood…or perhaps it was a need to "experience life" after suffering so many near-death experiences. But whatever the reasons, Crowley never stayed in one place for too long.

"My stepdad, Jerry, had started a painting business," he recalled. "He was a house painter and did pressure-washing and stuff. On weekends, I'd paint with him, and make some money that way." Looking for a long-term trade, however, Dan was accepted into the machinist program at Pratt & Whitney - a prominent aerospace manufacturer whose TF-30 engines were featured aboard the F-111 Aardvark and F-14A Tomcat. "They needed aircraft machinists."

The work was fascinating, but Dan preferred something where he wasn't always confined to a workshop. He then landed a job at the Palm Beach Post Times. "That was a pretty large newspaper in south Florida," he recalled. "I was a circulation manager there for about three and a half years." It wasn't long, however, before he felt the need to try something else.

"Then I started working for a pressure cleaning company." His new employer manufactured and sold pressure cleaners for homes and factories. And he developed a sincere fascination for many of the products he peddled. "We had one unit that could put out 10,000 psi. You could take a cement block and turn it into gravel!" he marveled. These high-powered units were for cleaning cement block factories, paint factories, power plants, and the like. "Then, I went to work for a signal company," he continued, "putting in traffic signals."

Dan then moved west to California, among the Redwood forests and vineyard valleys. He landed a job removing Redwood stumps from local vineyards—a task that leveraged the rubber tree-cutting and earthmoving skills he had honed in Vietnam. "Those Redwood stumps are huge," Dan recalled. He estimated that the trees themselves had been cut down in the 1850s, but the remaining stumps stood nearly five feet above the ground, and were nearly 20-30 feet across.

His next transitory job was a stint as a commercial fisherman before landing a semi-permanent job as a maintenance technician at a ski resort in Wyoming. Admittedly, it was a great gig for the wandering handyman. When he wasn't fixing ski lifts, he often hunted elk and went trout fishing at Lee Lake in the Grand Tetons. "That lake was just a mirror and you had the mountains in the background reflecting on the lake. It was one of the most picturesque things I've ever seen."

He stayed in Wyoming for the next several years until he felt the urge to return to Florida. "I went down to Key West," he said—the farthest territorial edge of America's Gulf Coast. "I got a really good job, doing maintenance at the Banyan Resort in downtown Key West." Around this time, Dan Crowley finally decided to set down roots in Florida. He got married and remained at the Banyan Resort for seven years before retiring for good. He and his wife, Trish, currently reside in Interlachen, Florida.

Jay Franz, after being discharged from the San Francisco hospital, returned to duty. His next assignment was a three-year tour as an ROTC Instructor at UC-Berkeley. Of course, in 1966, Berkeley was not yet the "hotbed of nonsense," it would become in later years. Indeed, as the war in Vietnam dragged on, Berkeley would gain notoriety for its association with protest groups like Students for a Democratic Society (SDS) and other elements of the 1960s counterculture.

Jay then reported to Thailand for a yearlong tour as part of the US support mission in Southeast Asia. His career then took an interesting turn as he was assigned to NORAD in Colorado Springs as a staff officer in the missile defense program. By 1970, however, NORAD was no longer the chic assignment it had been during the earlier days of the Cold War. Counterinsurgency and jungle warfare were dominating headlines (and most of the defense budget), so the traditional anti-Soviet missions appeared to be little more than an afterthought.

After an uneventful tour in Colorado Springs, Jay reported to the Defense Language Institute in Monterey, California. He would be enrolling in an intensive Spanish language course en route to his next assignment in the Panama Canal Zone. "I had a very interesting tour there," he said. "I was the assistant to the Commanding General of the US Southern Command." In this capacity, he worked for General George Mabry, a highly-decorated officer who had won the Medal of Honor during World War II.

Typically, US Army officers didn't get back-to-back overseas tours. Panama, despite being on the North American continent, was nevertheless considered an "overseas" posting by the Army's personnel managers. Thus, Jay did not expect an assignment to Naples, Italy when he requested it. But he happily accepted the position as an opportunity to work with NATO allies. "That was probably the most enjoyable assignment I've ever had in the service." He remained in Europe for the next several years, serving in a variety of US Army Europe postings.

By the 1980s, he had returned to the States, assuming a staff position in the Pentagon. After three years of cyclic staff work, however, Jay was ready for a change. His final assignment took him to Marietta, Georgia where he was an Active Duty Advisor to the Army National Guard. Jay retired in 1987 at the rank of Colonel, completing 28 years of active service. Upon his retirement from the Army, he launched a successful career in real estate. He retired for good in the early 2000s; today he resides in Tucson, Arizona.

After completing the Engineer Officer Advanced Course at Fort Belvoir, Larry Blair finally had the opportunity to attend graduate school. "Because I had been deferred a couple times already for graduate school, the Army said: 'You can take your pick of where you want to go to get your master's degree." Larry wanted to earn a Master of Science in Civil Engineering, but his schools of choice—The University of Illinois and the University of Missouri—were filled to capacity for the upcoming school term.

But this influx of student registrations was no coincidence.

"It turned out," said Larry, "that this was the last year of college deferments that they were giving out…so everybody was trying to get into graduate school to avoid the war." Thus, Larry found himself competing against latter-day draft dodgers. Undaunted, and determined to earn his graduate degree, Larry Blair asked the Army: "Can I go back to my alma mater? The South Dakota School of Mines?"

"Well, we don't have a program there…we don't have a

contract with them," the Army replied. "But if you can work it out with them, you can go there." A moment later, Larry was on the phone with a few of his old professors and the Dean of Students. "It turned out there was another guy…and Army captain in the Signal Corps who was doing the same thing I was," Larry continued. "So, the two of us were the first graduate students to go back to South Dakota under that Army program."

The proviso, however, was that he had to earn his Master's degree in *one* year.

"It was really tough," he admitted. "I had been out of school for eight years, so I was pretty rusty," he laughed. In fact, he had to attend summer school for "refresher courses" in calculus and thermodynamics. "Anyway, I managed to pass them, and then I got into the regular program…and I only had one year to do it. Normally, graduate students take at least a year and a half… usually two years."

But the Army told him: "You gotta do it in one year."

"So, we did it."

But because the South Dakota School of Mines had no prior liaison with the Army's graduate degree coordinators, the faculty was uncertain on how to handle Larry's degree plan.

"What do you want to do your thesis on?" they asked him.

"I don't really want to do a thesis," he said.

"I want to get as broad an education as I can in Civil Engineering, not focus on some particular little project."

The academic committee found his answer incomprehensible.

"Nobody's done that before," they said.

"Well, why can't I do it?"

The graduate advisors explained that students produced a "thesis" project, within a particular subset of Civil Engineering, under advisory of a specialist faculty committee. "They guide you through the process," he recalled, "and then gave you a comprehensive exam."

But Larry wasn't interested in doing that.

"I want to get as many courses as I can," he told them, "because I don't know what I'm going to be doing from now on." The Corps of Engineers could assign him to any variety of projects—so it made little sense to narrow his focus onto a particular academic track.

The school then offered him a compromise.

"Well, okay," they said, "but your committee is going to have to be a professor from *each* one of the disciplines that you *could* go into." Thus, Larry Blair took advanced courses in Concrete Structures, Hydraulics, and Geotechnical Engineering, among others. *Four* different professors sat on his final committee, administering an oral exam that lasted *eight hours*. "It was really interesting," he said, "with a lot of sweat and worry, but we made it through."

With his Master's degree in hand, Larry was certain that the Army would assign him to a nice stateside engineering job.

Instead, they sent him back to Vietnam.

For that second tour, "I had orders to the 8th Engineer Battalion of the 1st Cavalry Division," he recalled, but when he arrived in Vietnam, his orders inexplicably changed. "So, I ended up in the Engineer Section of the I Field Force."

By this time in the war, MACV had created two corps-level headquarters: I Field Force and II Field Force. "So, I stayed in that Engineer Section for a whole year [1969-70]. But the interesting part for me was that I got assigned to be the engineer advisor to the two engineer battalions of the South Korean Army that were there." The ROK had sent two of their best divisions to Vietnam. "And we paid their soldiers the same wages that are our US soldiers were getting," said Larry. "So, for a Korean soldier, it was a plum job."

Still, the ROK soldiers were very diligent in their work.

"They wouldn't screw up," said Larry, "because if they did, they got sent back to Korea. So, they would behave themselves. They were terrific troops."

After his second tour in Vietnam, Larry attended the Command and General Staff College at Fort Leavenworth, and then was assigned to Continental Army Command (CONARC) at Fort Monroe, Virginia. While he was there, CONARC split into two separate commands: Forces Command (FORSCOM) and the Training and Doctrine Command (TRADOC). FORSCOM oversaw the Army's mainline operational units, while TRADOC ran the Army's training centers. TRADOC itself was borne from the darkest days of Vietnam, when the Army realized it had to return to realistic field-based training, focusing on small-unit tactics and basic combat skills.

"I stayed at TRADOC for three years and then got assigned to Korea. I took the family over to Korea for two years, and my job there was the Army's Contracting Representative for the maintenance contracts on all the Army bases." At the time, the Army didn't have any organic maintenance crews for their bases in Korea. Thus, to maintain their various posts across the DMZ, the Army out-sourced regular maintenance to a few private South Korean firms. "My job was to make sure that those contractors were doing their jobs." He had to account for their work orders and the associated budgets. "I got to travel all over Korea…and I had a Korean counterpart, a Korean civil engineer. He went with me everywhere because I couldn't speak Korean, and he could translate."

Coming home from Korea, Larry Blair was reassigned to Albuquerque, New Mexico. "I was the Deputy District Engineer for the Albuquerque District Corps of Engineers," he said. "And I spent four years in that job, partly because the District Engineer retired, and they wanted me to provide some continuity between the old one and the new one. So, I got extended for a year."

As Larry recalled, the Albuquerque posting was a "plum job." Most of the staff were civilians, and the work hours were generally stable. "Our district went from West Texas to all of New Mexico, Southern Colorado, and western Kansas. And we were doing flood control work and all over the district.

Larry's final assignment was to Fort Carson, Colorado as the Post Engineer. "I spent my last two years at Fort Carson." He retired from active duty in 1981 at the rank of Lieutenant Colonel. Having spent a few years in Albuquerque, however, he was determined to re-settle in the American Southwest. As luck would have it, he secured a job with the Flood Control Authority in Albuquerque—"and I ended up working for the same Colonel I had worked for before [in the Corps of Engineers] who had also retired." He spent the next sixteen years with the Flood Control Authority before being named the Director cf Public Works for the City of Albuquerque. He retired for good in 2001. He and his family still reside in Albuquerque, New Mexico.

Although Chuck Humphrey resigned his Regular Army commission following Vietnam, he elected to stay in the Army Reserve. He arrived home in Fargo, North Dakota on September 30, 1966. "I had a pretty good homecoming," he recalled. It was still early in the war, and the public backlash hadn't erupted yet. "My mother, my sister, my wife, and her parents were waiting for me."

Chuck had a near-seamless transition into civilian life. Shortly after returning home, he landed a job working for the US Department of Agriculture. In January 1967, he received orders to the local Army Reserve engineer company in Fargo. "And I stayed in that unit in Fargo, and actually commanded it for a couple of years. I stayed until there until 1970." That year, he elected to pursue a PhD in "Adult Education." With that, he moved his family from Fargo, North Dakota to Madison, Wisconsin where completed his doctoral studies at the University of Wisconsin (UW). All the while, he was fortunate to have escaped the wrath of American protestors and the VVAW. "I came home in late '66," he said, "and the dissatisfaction of the American public with the Vietnam War hadn't really set in yet."

However, by the time he arrived at UW-Madison, the campus had become an unlikely "ground zero" in the anti-war movement. On October 18, 1967, UW students protested the on-campus recruiting efforts of Dow Chemical—a company that manufactured napalm. In February 1969, the Black Student Strike brought campus activities to a standstill. However, in August 1970, UW suffered one of the worst domestic terrorist acts in American history. Four disaffected young men loaded a Ford Econoline van with ammonium nitrate fertilizer, and parked it next to UW's Sterling Hall, which housed the Army's Mathematics Research Center. When the truck bomb detonated, it killed one researcher, "and knocked out windows for about 20 blocks in every direction," said Chuck. "So, by then, I had become aware of the unpopularity of the war."

But despite the explosions and protests happening around him, Chuck was fortunate to have never received any personally-directed vitriol. "I am not one of those veterans that ever got spat upon and sworn at, and all the other things you've heard about… particularly enlisted soldiers who were released from active duty, and came home in that period between 1967-73." In fact, Chuck had been a keynote speaker at several Memorial Day services throughout the 1970s, and he "never experienced anything other than respect for being a member of the US Army, a reservist, and now a retired reservist." Still, he admits that this experience was not typical. He was one of the exceptions; not the rule.

Chuck remained in the Army Reserve until 1992, retiring at the rank of Lieutenant Colonel. Upon finishing his PhD in 1973, Chuck began his career as a college professor, teaching for several years at Boise State University in Idaho. Today he is semi-retired, still giving occasional lectures at the university as a professor emeritus. He and his wife, Jane, currently reside in Pocatello, Idaho.

ABOUT THE AUTHOR

MIKE GUARDIA is an internationally-recognized author and military historian. A veteran of the United States Army, he served six years on active duty as an Armor Officer. He is the author of the widely-acclaimed *Hal Moore: A Soldier Once…and Always*, the first-ever biography chronicling the life of LTG Harold G. Moore, whose battlefield leadership was popularized by the film *We Were Soldiers*, starring Mel Gibson.

He was named "Author of the Year" in 2021 by the Military Writers Society of America, and has been nominated twice for the Army Historical Foundation's Distinguished Book Award.

As a speaker, he has given presentations at the US Special Operations Command, the George HW Bush Presidential Library, the First Division Museum, and the US 7th Infantry Division Headquarters at Fort Lewis.

In 2022, he appeared in the History Channel series, *I Was There*, cast as a featured historian in the episodes on the Johnstown Flood of 1889, the Chernobyl Disaster, the Battle of Stalingrad, and the Oklahoma City Bombing. His other media appearances include guest spots on *National Public Radio* (NPR); *Frontlines of Freedom*; *Armada International*; and *Military Network Radio*.

His work has been reviewed in the *Washington Times*, *Military Review*, *Vietnam Magazine*, *DefenceWeb South Africa*, and *Soldier Magazine UK*. He holds a BA and MA in American History from the University of Houston; and an MA in Education from the University of St. Thomas. He currently lives in Minnesota.

SELECT BIBLIOGRAPHY

Interviews

Interview with Larry Blair; February 22, 2022.

Interview with Larry Blair; April 7, 2022.

Interview with Larry Blair; May 6, 2022.

Interview with Larry Blair; July 17, 2022.

Interview with Dan Crowley; February 19, 2022.

Interview with Dan Crowley; March 17, 2022.

Interview with Dan Crowley; April 5, 2022.

Interview with Dan Crowley; June 17, 2022.

Interview with Dan Crowley; July 5, 2022.

Interview with Dan Crowley; September 18, 2022.

Interview with Jay Franz; April 14, 2022.

Interview with Jay Franz; May 4, 2022.

Interview with Chuck Humphrey; October 21, 2022.

Interview with Chuck Humphrey; November 3, 2022.

Interview with Chuck Humphrey; December 11, 2022.

Primary Sources

The Daniel Crowley Papers. A collection of Crowley's various personal papers, photographs, operational reports and the oral history covering his military career. Also included are eighty-five (85) pages of typed notes and narratives, spiral-bound, interspersed with various news magazine articles covering the major operations in Vietnam, 1965-66.

The Fowler C. Humphrey Photo Collection. A collection of various photographs covering the 1st Engineer Battalion's deployment to Vietnam, 1965-66.

Sheehan, Neil; et al. *The Pentagon Papers: The Secret History of the Vietnam War.* Skyhorse: New York, 2017.

US Army Corps of Engineers. *In Vietnam: A Pictorial History of the 1st Engineer Battalion, 1st Infantry Division, October 1965 – March 1967.* Government Printing Office: Fort Riley, 1967.

United States Army. *The First Infantry Division in Vietnam: 1965-1967.* Office of Information and Public Affairs: APO San Franscisco, 1967.

Secondary Sources:

Carland, John. *Combat Operations: Stemming the Tide, May 1965-October 1966.* (US Army in Vietnam Series). US Army Center for Military History: Washington DC, 2000.

Cosmas, Graham. *MACV: The Joint Command in the Years of Withdrawal, 1968-1973.* (US Army in Vietnam Series). US Army Center for Military History: Washington DC, 2008.

Guardia, Mike. *Danger Forward: The Forgotten Wars of General Paul F. Gorman.* Magnum Books: Maple Grove, 2021.

Guardia, Mike. *Hal Moore: A Soldier Once…and Always.* Casemate: Havertown, 2013.

Ploger, Robert. *US Army Engineers, 1965-1970* (Vietnam Studies). US Army Center for Military History: Washington DC, 1974.

Traas, Adrian G. *Engineers at War* (US Army in Vietnam Series). US Army Center for Military History: Washington DC, 2010.